T0135660

Chemical Orientation
of European Cockchafers,
Melolontha melolontha L.

Dissertation zur Erlangung
des akademischen Grades des
Doktors der Naturwissenschaften
(Dr. rer. nat.)

eingereicht im Fachbereich
Biologie, Chemie, Pharmazie der
Freien Universität Berlin

von

Andreas Reinecke

im Dezember 2005

Bibliographic information published by Die Deutsche Bibliothek

Die Deutsche Bibliothek lists this publication in the
Deutsche Nationalbibliografie; detailed bibliographic data
is available in the Internet at http://dnb.ddb.de.

cover photo by: Andreas Reinecke

ISBN 3-8325-1205-5

Logos Verlag Berlin
Comeniushof, Gubener Str. 47,
10243 Berlin
Tel.: +49 030 42 85 10 90
Fax: +49 030 42 85 10 92
INTERNET: http://www.logos-verlag.de

Die Dissertation wurde am Institut für Biologie der Freien Universität Berlin in der Arbeitsgruppe Angewandte Zoologie/Ökologie der Tiere unter der Anleitung von Prof. Dr. Monika Hilker in enger Kooperation mit PD Dr. Joachim Ruther angefertigt.

1. Gutachterin: Prof. Dr. Monika Hilker

2. Gutachter: Prof. Dr. Gerd Weigmann

Die Disputation erfolgte am 20. Dezember 2005.

This thesis is based on the following manuscripts:

Andreas Reinecke, Joachim Ruther & Monika Hilker (2002). The scent of food and defence: Green leaf volatiles and toluquinone as sex attractant mediate mate finding in the European cockchafer *Melolontha melolontha*. **Ecology Letters** 5:257-263.

Joachim Ruther, Andreas Reinecke, Till Tolasch & Monika Hilker (2002). Phenol – Another cockchafer attractant shared by *Melolontha hippocastani* and *M. melolontha*. **Zeitschrift für Naturforschung** C 57:910-913.

Andreas Reinecke, Joachim Ruther & Monika Hilker. Precopulatory isolation in sympatric *Melolontha* species? Submitted to **Agricultural and Forest Entomology**.

Andreas Reinecke, Joachim Ruther, Till Tolasch, Wittko Francke & Monika Hilker (2002). Alcoholism in cockchafers: Orientation of male *Melolontha melolontha* towards green leaf alcohols. **Naturwissenschaften** 89:265-269.

Andreas Reinecke, Joachim Ruther & Monika Hilker (2005). Electrophysiological and behavioural response of *Melolontha melolontha* to saturated and unsaturated aliphatic alcohols. **Entomologia Experimentalis et Applicata** 115:33-40.

Andreas Reinecke, Joachim Ruther, Christoph J. Mayer & Monika Hilker (2006). Optimized trap lure for *Melolontha* species. **Journal of Applied Entomology** in press.

Andreas Reinecke, Frank Müller & Monika Hilker. Do plant roots hide? Attractiveness of CO_2 to white grubs fades on the background of root exudates. Manuscript.

References are listed at the end of each manuscript based chapter (Chap. 2-8). A combined reference list for Chap. 1, 'Introduction', and Chap. 9,'General discussion', is placed at the end of Chap. 9.

Contents

List of Figures

List of Tables

Chapter 1

Introduction

Important damages in agriculture and horticulture have been a driving force to study the chemical ecology of phytophagous scarab beetles during the past decades (Leal 1998). Pheromones and plant-derived semiochemicals may serve in pest insect control or monitoring.

Scarab beetle aggregation and sex pheromones are as diverse as aliphatic aldehydes, ketones, alcohols, or esters, lactones, aromatic compounds, amino acid derivatives, terpenoids, and alkaloids (review in Leal 1998; Rochat *et al.* 2002; Ward *et al.* 2002; Leal *et al.* 2003; Nojima *et al.* 2003a,b; Tolasch *et al.* 2003; Tóth *et al.* 2003). As in other insects, plant volatiles may enhance the attractiveness of pheromones (Landolt & Philips 1997; Leal 1998; Reddy & Guerrero 2004).

Intriguingly, perception of ubiquitous leaf volatiles by scarab beetle olfactory receptor neurones may be as selective and as sensitive as pheromone perception by moth receptor neurons (Hansson *et al.* 1999; Larsson *et al.* 2001; Stensmyr *et al.* 2001). However, the ecological function of these highly sensitive neurons remained unclear, until Ruther *et al.* (2000) described the mate finding strategy of *Melolontha hippocastani* FABR.[1], which is based on feeding-induced plant volatiles as primary sex attractants.

Further investigations revealed that 1,4-benzoquinone, which is known as defensive compound in other insect species, functions as sex pheromone in this species, adding a new group of chemicals to the list of scarab beetles sex pheromones (Ruther *et al.* 2001). It may be surprising that chemical orientation has not earlier been investigated in this important forest pest. But it has been even more surprising that, until this study started, hardly anything was known about the chemical ecology of *Melolontha melolontha* L.[2], the other species of the genus, which is common and a severe pest in agriculture as well as horticulture in central Europe.

A long history of calamitous mass breeding incidents is documented for *M. melolontha* (Keller 1986a). The earliest reports on pest control actions taken against cockchafers date back to 1478/1479, when the bishop of Lausanne organized canonical lawsuits against this insect pest. Beginning by the end of the 18th century, more scientific methods of pest control were implemented (Keller 1986). Concurrently, intensive fundamental research regarding ecology, behaviour, physiology, and morphology of cockchafers led to numerous publications, which have been summarized and extended by prominent authors in the field (Zweigelt 1928; Schwerdtfeger 1939; Ene 1942; Hurpin 1962; Niklas 1974; Keller 1986a, 1986b; Krell 2004). Although some authors have considered chemical ecological aspects, experimental evidence for

---

[1] *Melolontha hippocastani* FABRICIUS (1801); „forest cockchafer"; (Coleptera: Scarabaeoidea: Scarabaeidae: Melolonthinae).

[2] *Melolontha melolontha* LINNAEUS (1758); syn. *M. vulgaris* FABRICIUS; syn. *Hoplosternus melolontha* L.; syn. *Scarabaeus majalis* MOLL; syn. *Scarabaeus melolontha* L.; „European" or „common cockchafer".

behavioural responses of adult beetles to olfactory stimuli was lacking. Instead, accounts about putative chemical stimuli, reviewed by Krell and Fery (1992), and the assumption of attractive plant volatiles by Niklas (1974), were reported in the literature.

Most of the economic damages caused by cockchafers are attributable to the root feeding larvae. The orientation of the so-called white grubs to plant roots has experimentally been investigated by Ene (1942), Klingler (1956), and Hasler (1986). Carbon dioxide, an ubiquitous primary metabolite, is so far the only identified attractant guiding *M. melolontha* larvae to host plant roots (Hasler 1986). However, pronounced feeding preferences, which even differ between larval stages (Ene 1942; Hauss 1975; Hauss & Schütte 1976; Keller 1986b), indicate that localizing preferred roots from a distance might be an advantage to white grubs.

As Karg and Suckling pointed out (1999), application of semiochemicals in pest control needs thorough fundamental understanding of a target insect's behaviour and the mechanisms underlying the desired behavioural modifications. Not every new insight necessarily leads to applications. However, scientific knowledge is an indispensable prerequisite for successful and sustainable pest control. To promote both, fundamental ecological understanding and environmental sound means of pest control, chemical orientation of *M. melolontha* adults above and larvae below ground has been investigated.

1.1 Above ground communication and orientation

Olfaction in insects serves both intra-specific communication and the location of hosts. Since the first identification of a lepidopteran pheromone by Butenandt *et al.* (1959), pest control techniques applying pheromones and their synthetic analogues have been developed and partially replaced the use of insecticides (Karg & Suckling 1999; Schlyter & Birgersson 1999; Renou & Guerrero 2000). Although many scarab species, especially Rutelinae and Melolonthinae, are severe pests, monitoring and control techniques were developed on a try and error basis using (host-) plant volatiles as lures (Leal 1998). It took until 1970 to identify, rather by chance, the first scarab beetle sex pheromone. Male *Costelytra zealandica* WHITE beetles were attracted to Pliobond glue, which contains phenol (Henzell & Lowe 1970). Only two more scarab sex pheromones were identified until the end of the 1980ies (Tumlinson *et al.* 1977; Tamaki *et al.* 1985), but many more followed since then (Leal 1998; Ruther *et al.* 2001; Reinecke *et al.* 2002; Rochat *et al.* 2002; Ward *et al.* 2002; Leal *et al.* 2003; Nojima *et al.* 2003a; Tolasch *et al.* 2003; Tóth *et al.* 2003). Synergistic or additive interactions between scarab pheromones as primary sex attractants and olfactory host plant stimuli were, as in other insects, recognized soon after pheromone identification (Landolt & Philips 1997; Leal 1998; Reddy & Guerrero 2004). In Melolonthinae and Rutelinae known sex pheromone communication follows exclusively the female emitter – male receiver scheme.

The elucidation of the *M. hippocastani* mate finding strategy added new facets to the knowledge on chemical communication in insects. The process of mate finding is based on a sexually dimorph swarming flight behaviour. As primary sex attractant, a green leaf volatile (Chap. 4, Chap. 5), emitted by host tree leaves upon female feeding, guides flying males to females remaining on the host leaves. The notion SEXUAL

KAIROMONE has been introduced into the scientific discussion to describe the function of the plant volatile. A female-derived sex pheromone synergistically enhances the attractiveness of the leaf volatile to males (Ruther *et al.* 2000, 2001, 2002a, 2003).

Based on the previously acquired knowledge in *M. hippocastani*, olfactory orientation of *M. melolontha* adults and its function in host and mate finding was investigated addressing the following questions.

1.1.1 Mate finding and pheromone related communication

- *Chapter 2: Sex pheromone identification.*
  How do *M. melolontha* males find mates? Is a sex pheromone involved? And if so, which is the chemical structure, and how does it interact with plant volatiles?

- *Chapter 3: Other beetle-derived male attracting compounds.*
  Are other beetle-derived compounds involved in mate finding? How do they interact with each other and with plant volatiles?

- *Chapter 4: Precopulatory species isolation in the genus* Melolontha.
  M. melolontha and *M. hippocastani* co-occur in some areas. Do both species display the same sexual dimorphism in flight behaviour? Do males discriminate between conspecific and hetero-specific females based on olfaction? Which conclusion can be drawn from these findings regarding the pheromone communication in both species?

1.1.2 Host plant volatiles

- *Chapter 5: Function of host plant volatiles.*
  Are plant volatiles involved in host or mate finding? Do intact or damaged host plant leaves attract *M. melolontha* beetles? Which leaf volatiles do the beetles' antennae perceive? Which compounds are most attractive in the field? Are both sexes equally attracted to host leaf volatiles?

- *Chapter 6: Structure-activity relations.*
  Leaf alcohols attract male cockchafers in the field (Chap. 5). Does perception and attractiveness of leaf alcohols and functional analogues depend on the structure of the molecule? Do male and female antennae generate equal dose-response profiles when stimulated with these compounds?

1.1.3 Application aspects

- *Chapter 7: Optimized trap catches.*
  M. melolontha males are attracted to beetle-derived compounds, host leaf volatiles, and mixtures thereof as described in Chap. 2, Chap. 3, Chap. 5, and Chap. 6. Do special mixing ratios optimize trap catches?

1.2 Below ground orientation

In sharp contrast to the wealth of knowledge available on chemical communication and orientation of insect live stages living above ground, information about host cues exploited by herbivorous arthropods below ground is scarce. Other organisms, as diverse as root associated bacteria, mycorrhizal fungi, or phytoparasitic nematodes, orientate by using root exudates or root volatiles from intact plants (e.g. Mateille 1994; Zheng & Sinclair 1996; Vierheilig 1998). In a recent review on soil living herbivorous insects, Johnson and Gregory (2006) reported 14 species being attracted to carbon dioxide (CO_2) sources, among them two dipteran, one lepidopteran, and 11 coleopteran species. Eight insect species use other plant-derived compounds to find host plant roots. Larvae of a root weevil locate their host, a Leguminosae, by a compound concentrated in nitrogen-fixing root nodules (Johnson *et al.* 2005). Others, among them a scarab species, are known to be attracted to specific roots, but the attractive compounds have not been identified (Johnson & Gregory 2006). In tritrophic systems, plants roots have recently been shown to attract entomopathogenic nematodes after herbivore attack (van Tol *et al.* 2001; Rasmann *et al.* 2005).

Although generalists, *Melolontha* larvae show distinct feeding preferences. Thus, successful finding of preferred roots and the ability to discriminate them from others might be an evolutionary advantage (see above). In interactions between root-feeding insect larvae and their hosts both foraging kairomones (Ruther *et al.* 2002b), beneficial to the herbivore, and allomones (Dicke & Sabelis 1988), beneficial to the plant, have to be considered. Earlier experimental setups were limiting in central aspects like reduced root - rhizosphere interface, the use of damaged root material, and the exclusion of water-soluble compounds from the bioassays (see Chap. 7). Hence, a compartmented soil arena was designed to investigate *M. melolontha* white grub responses to stimuli from intact plant roots.

1.2.1 Oriented responses below ground

- *Chapter 8: Localization of host plant roots.*
  Do cockchafer larvae orient to the known attractant CO_2 in the newly designed arena? Are intact plant roots attractive to white grubs? Do root-derived chemicals interact with CO_2 when presented as extracted exudates?

Chapter 2

The scent of food and defence: Green leaf volatiles and toluquinone as sex attractant mediate mate finding in the European cockchafer *Melolontha melolontha*

Andreas Reinecke, Joachim Ruther, and Monika Hilker

Ecology Letters (2002) **5**: 257-263 [1]

Abstract:
Mate finding in the European cockchafer, *Melolontha melolontha*, has been investigated in the field. Our data suggest that mainly males show flight activity at dusk. In a landing cage bioassay, male cockchafers preferred cages baited with females over cages baited with males. Gas chromatographic analysis of beetle extracts with electronantennographic detection revealed the presence of electrophysiologically active compounds, among them toluquinone. In funnel trap bioassays none of these compounds was attractive towards males *per se*. A mixture of green leaf volatiles (GLV) mimicking the bouquet of mechanically damaged leaves was highly attractive to *M. melolontha* males. The attractiveness of the same GLV mixture was synergistically enhanced when toluquinone was added to the lure. GLV attract males to damaged leaves. Toluquinone as a sex pheromone indicates the presence of conspecific females and synergistically enhances the attractiveness of GLV.

Keywords: GC-EAD, kairomone, *Melolontha melolontha*, plant volatiles, Scarabaeidae, sex pheromone.

2.1 Introduction

In Central Europe mass outbreaks of the European cockchafer (maybeetles), *Melolontha melolontha* L. (Coleoptera: Scarabaeidae), are known every 30–40 years (Keller 1986). While egg deposition and larval development take place in the soil of open fields and orchards, adult beetles gather on forest edges to feed and mate. Depending on the local climate, the life cycle of *M. melolontha* is commonly 3–4 years long, with a short period of adult emergence from the middle of April through to the end of May (Hurpin 1962; Keller 1993). Even though *M. melolontha* is one of the best known insects in central Europe, hardly anything is known about its mate finding.

Various sex pheromones and some aggregation pheromones have been identified in other phytophagous scarab beetles, among them fatty acid derivatives, terpenoids, amino acid derivatives, aromatic compounds and an aromatic alkaloid (Hallett *et al.* 1995; Leal 1998). Plant volatiles may synergistically enhance the attractiveness of pheromone lures (Klein *et al.* 1981; Leal *et al.* 1994a; Yarden & Shani 1994). However, all so-far identified major compounds of scarab sex pheromones were attractive *per se* to conspecifics of the opposite sex (e.g. Tumlinson *et al.* 1977; Tamaki *et al.* 1985; Leal *et al.* 1992a,b, 1993a,b, 1994b; Leal 1993; Facundo *et al.* 1994; Zhang *et al.* 1994, 1996; Hallet *et al.* 1995; Bauernfeind *et al.* 1999; Kakizaki *et al.* 2000; Ruther *et al.* 2001a).

Males of the forest cockchafer, *Melolontha hippocastani* FABR., a close relative of *M. melolontha*, have recently been shown to be attracted by green leaf volatiles and a sex pheromone (Ruther *et al.* 2000). The sex pheromone that has been identified as 1,4-benzoquinone is slightly attractive *per se* but synergistically enhances the response of males towards green leaf volatiles (Ruther *et al.* 2001a). Thus, a novel mechanism of mate finding has been proposed for *M. hippocastani* (Ruther *et al.* 2001a). Initially, males are attracted towards damage-induced green leaf volatiles released by leaves that females feed upon. The sex pheromone allows for discrimination of unspecific leaf damage and female feeding sites.

Males of *M. melolontha* were found to be attracted by the green leaf alcohols 1-hexanol, (*E*)-2-hexenol, and (*Z*)-3-hexen-1-ol (Reinecke *et al.* 2002). Here, we report the identification of the sex pheromone of this species. In contrast to so-far identified sex pheromones of scarab beetles, the active compound is not attractive *per se*, but synergizes the male response towards green leaf volatiles.

2.2 Methods and materials

2.2.1 Chemical analyses and electrophysiological experiments

The production of a sex pheromone in scarab beetles may drop after mating (Leal *et al.* 1993b). Both male and female cockchafers mate several times from the emergence flight onwards (Krell 1996). Therefore, only unmated beetles were used for bioassays and chemical analysis. They were dug up in an apple plantation (Obergrombach, Baden-Württemberg) prior to the flight season, sexed and kept in insect cages under natural conditions until needed. Fresh leaves from *Carpinus betulus* L., *Fagus sylvatica* L., and *Quercus robur* L. were fed *ad libitum*. Full body extraction started at the beginning of the swarming period (at about 21.00), when mating activity increases in the field. Males and females remained for 20 hr in 2 mL dichloromethane per beetle. Raw extracts were dried over anhydrous sodium sulphate. Ten male or female equivalents (20 mL) were concentrated under a gentle stream of nitrogen. At the beginning of the concentration process, 50 ng methyl benzoate in 5 μL dichloromethane were added as an internal standard. The concentrate was fractionated using a 100-mg silica gel cartridge for solid phase extraction (IST, Mid-Glamorgan, UK). The cartridge was solvated with 1 mL pentane. After applying the sample to the cartridge it was stepwise eluted with 1 mL of (a) pentane, (b) 10% dichloromethane in pentane, (c) dichloromethane and (d) methanol. The fractions were carefully concentrated to 20 μL under a gentle stream of nitrogen and used for chemical analysis.

Gas chromatography with electroantennographic detection (GC-EAD)

Electrophysiological responses of male *M. melolontha* towards compounds from extracts of males and females were recorded by gas chromatography with electroantennographic detection in order to locate pheromone candidate substances: 1–2 μL of the fractionated beetle extracts were injected in splitless mode (1 min, injector temperature 240°C) onto a fused silica column (30 $*$ 0.32 mm ID) coated with DB-5 stationary phase (film thickness 0.25 μm) of a Fisons 8060 gas chromatograph (Thermoquest, Egelsbach, Germany). The oven temperature was 40°C during injection of the sample and programmed to rise immediately at a rate of 10°C min^{-1} to a maximum of 280°C. Helium was used as carrier gas at a flow rate of 2 mL min^{-1}. The injector temperature was kept at 240°C. The effluent of the GC-column was split into two equal parts. One part was transferred to a flame ionization detector, the other part via a heated transfer line (225°C) to an electroantennographic detector (Syntech, Hilversum, The Netherlands). After dissection of *Melolontha* antennae close to the base, they were fixed by use of a conductive gel (Spectra 300 electrode Gel, Parker Laboratories, Orange, New Jersey, USA) on a dual electrode MTP-4 probe (Syntech). The antennal pedicel was fixed on one electrode, the tip of the last antennal lamella on the other. In order to optimize the air flow close to the sensillae, antennal lamellae were separated from each other as far as possible between the electrodes.

The antennal preparation was connected to a signal box (Autospike IDAC Box, Syntech) and inserted into a glass tube (0.7 cm ID) carrying the GC-effluents in a stream of cleaned air (1.5 L min^{-1}) by the aid of a micromanipulator. Antennal signals were recorded on a personal computer using the Electro Antennographic Detection software (Syntech). A mixture of synthetic 1,4-benzoquinone, toluquinone and nonanal at a concentration of 1 mg mL^{-1} per compound was analysed by gas chromatography with electroantennographic detection to confirm the electrophysiological activity of these pheromone candidates. (*Z*)-3-hexen-1-ol was added to the mixture as a positive control to test for responsiveness of the antennae. This compound has been found to elicit high responses on male and female antennae of *M. melolontha* (Reinecke *et al.* 2002). Before and after each GC-EAD run the antennal function was checked by two puffs of clean air loaded with the solvent alone or with (*Z*)-3-hexen-1-ol diluted in dichloromethane (10 mg mL^{-1}). For this purpose, 1 μL of either solvent or (*Z*)-3-hexen-1-ol diluted in dichloromethane was applied on a small paper disk (r = 5 mm) fitted into a Pasteur pipette. A first puff of 0.5 s duration at a flow rate of 25 mL s^{-1} blew off most of the solvent. After insertion of the Pasteur pipette into a small hole in the glass tube carrying the GC-effluent, a second puff with the same parameters was used to test the responsiveness of the antennae.

Gas chromatography with coupled mass spectrometry (GC-MS)

Fractionated beetle extracts (1–$2\,\mu$L) were injected in splitless mode ($1\,$min) onto the same type of fused silica column as described for the gas chromatography with electroantennographic detection. The oven temperature was kept at $40°$C for $4\,$min and programmed to rise at $10°$C min^{-1} to $280°$C. All other chromatographic conditions were the same as described above. The column effluent was transferred into a coupled Fisons MD 800 quadrupole mass spectrometer (Thermoquest) and ionized by electron impact ionization at $70\,$eV. Identification of the compounds was based on comparison of mass spectra and retention times with those of authentic reference compounds.

2.2.2 Field trials

Exp. 1 and 3 were performed between 29 April and 8 May 2001 at the edge of forests close to Obergrombach, Baden-Württemberg. The stand of trees in the experimental area was dominated by *F. sylvatica*, whereas the forest edges were intermixed with *Q. robur*, *Quercus petraea* LIEBL., *C. betulus* and *Acer pseudoplatanus* L. Exp. 2 was performed between 8 and 10 May 2000 in a *Q. robur* stand in Niederbeerbach, Hessen that was heavily infested by *M. melolontha*.

Landing cage experiment: Male response to conspecific males and females

Experiment 1. The landing cage bioassay has been developed to observe the response of flying *M. hippocastani* males towards conspecifics under field conditions (Ruther *et al.* 2000). Wire mesh cages ($160 * 70 * 40\,$mm, mesh $5\,$mm) were mounted pairwise on twigs of infested host trees (*C. betulus*, *F. sylvatica*) at a distance of 0.5–$1\,$m. Cages were baited either with five unmated females or five unmated males. Beetles within the cages were allowed to feed on the foliage of the twigs. Positions of test and control cages were randomized within each pair. Landing cages were baited $30\,$min before onset of the flight period. During the swarming flight, landings of beetles were counted for $30\,$min. Landing beetles were sexed, removed from the landing cage and kept isolated until the end of the experiment to prevent double counting. Numbers of landings on the cages were statistically compared by the Wilcoxon matched pairs test (Sachs 1992) ($n = 15$ replications).

Funnel trap bioassays. Test for attractiveness of 1,4-benzoquinone and toluquinone

Funnel traps were used as described by Ruther *et al.* (2000). They consisted of 2-L polyethylene wide-mouth bottles and 185-mm powder funnels (outlet diameter $40\,$mm) equipped with four crosswise arranged plastic crashing-sheets towering $100\,$mm above the funnel. Round boxes (ID $* $ H: $36 * 29\,$mm) were fixed on top of the crashing sheets. Lure substances and mixtures were dissolved in $500\,\mu$L dichloromethane, applied on balls of cotton wool, and placed into the round boxes. Traps were arranged in a complete block design with each block consisting of baited traps and a solvent control trap. Traps were randomized and placed at equivalent positions into one tree per block. Height, distance from the stem, light conditions and development of the foliage were almost identical. The distance between traps within a block was at least 1 m. Traps were baited and placed into the trees about 1.5 hr before the flight period. Catches were counted and sexed the next morning. Numbers of beetles caught in the differently baited traps within the different blocks were analysed using the non-parametric Friedman-ANOVA for dependent data followed by Bonferroni-corrected multiple Wilcoxon matched pairs tests (Sachs 1992).

Experiment 2. Each block consisted of four traps baited with (a) 5 mg 1,4-benzoquinone, (b) 5 mg of a green leaf volatile mixture composed of (Z)-3-hexenyl acetate (3.2 mg), (Z)-3-hexen-1-ol (1.5 mg), (E)-2-hexenal (0.2 mg), and (Z)-3-hexenal (0.1 mg), (c) 5 mg of 1,4-benzoquinone and 5 mg of the green leaf volatile mixture described in (b), and (d) 500 μL dichloromethane as a solvent control ($n = 15$ replications). The green leaf volatile mixture was the same as used by Ruther *et al.* (2001a, 2002).

Experiment 3. Traps were baited as follows: (a) 5 mg 2-methyl 1,4-benzoquinone (toluquinone), (b) 5 mg of the green leaf volatile mixture 2, (c) 5 mg of toluquinone and 5 mg of the green leaf volatile mixture mentioned above, and (d) 500 μL dichloromethane as a solvent control ($n = 21$ replications).

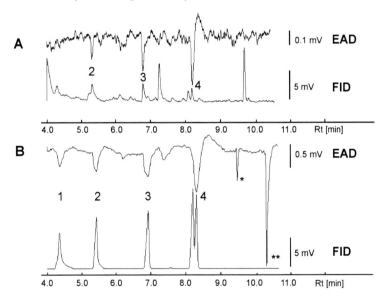

Figure 2.1: Gas chromatogram with simultaneous flame ionization detection (FID) and electroantennographic detection (EAD) using a male antenna of *M. melolontha*. (A) Analysis of 1 μL of a female extract representing 0.5 female equivalents. (B) Analysis of a synthetic mixture containing 1 μg/μL each of (Z)-3-hexen-1-ol (1), 1,4-benzoquinone (2), toluquinone (3), and nonanal (4). * response towards a test puff with 1 μL of the solvent dichloromethane, ** response towards a test puff with 1 μL (Z)-3-hexen-1-ol diluted in dichloromethane ($10\,\mathrm{mg\,mL}^{-1}$).

2.3 Results

2.3.1 Identification of potential pheromone compounds

Testing male antennae, electrophysiologically active compounds were detected in the dichloromethane fraction from female extracts by the use of gas chromatography with electroantennographic detection (Fig. 2.1 A). Among these compounds we identified 1,4-benzoquinone, toluquinone and nonanal using gas chromatography with coupled mass spectrometry. Synthetic reference compounds led to electroantennographic responses equivalent to those elicited by the beetle-derived chemicals, confirming the identity of the physiologically active compounds (Fig. 2.1 B).

Furthermore, 1,4-benzoquinone, toluquinone and nonanal were detected also in male extracts. When testing the response of female antennae towards these compounds by GC-EAD, each compound was found to elicit responses comparable with responses of male antennae.

2.3.2 Field trials

Landing cage experiment

During 15 replicates, 34 landings of male *M. melolontha* were counted on cages baited with male beetles and 158 male landings on cages with female beetles (Fig. 2.2). Only two female landings on cages baited with male beetles were observed and none on cages with female beetles.

Funnel trap experiments

A total of 567 male and 12 female cockchafers was caught in Exp. 2 and 3. The overall flight activity varied widely. Only 150 males were trapped in Exp. 2, while 417 male and 12 female catches were counted in Exp. 3. The small number of female catches and their distribution without detectable preference for any treatment did not allow statistical comparisons between treatments. Thus, only numbers of male catches were used in statistical analysis.

Experiment 2: Test for attractiveness of 1,4-benzoquinone. Traps baited with 1,4-benzoquinone alone did not catch significantly more male cockchafers (eight) than empty control traps (three). Traps baited with a mixture of synthetic green leaf volatiles (66) or a combination of the synthetic green leaf volatile mixture and 1,4-benzoquinone (73) caught significantly more males than empty controls or 1,4-benzoquinone-baited traps. However, numbers

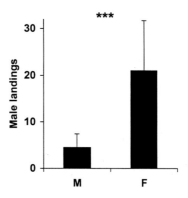

Figure 2.2: Response of male *M. melolontha* towards conspecifics in the landing cage experiment. Mean number of males ($\pm$ SD) landing during a 30-min observation period on cages baited with either male (M) or female (F) beetles having access to foliage. Wilcoxon matched pairs test, ***significance at $P < 0.001$, $n = 15$.

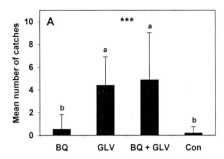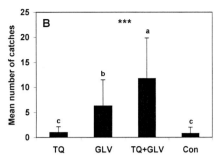

Figure 2.3: Response of *M. melolontha* males in funnel trap experiments (mean number of catches $\pm$ SD) towards green leaf volatiles in combination with (A) 1,4-benzoquinone and (B) toluquinone during the swarming period at dusk. Con: solvent control dichloromethane; BQ: 5 mg of 1,4-benzoquinone; TQ: 5 mg of toluquinone; GLV: 5 mg of synthetic green leaf volatiles; ***significance at $P < 0.001$ (Friedmann ANOVA) with (A) $n = 15$, df $= 3$, $\chi^2 = 37.35$, and (B) $n = 21$, df $= 3$, $\chi^2 = 50.50$; columns with different lowercase letters are significantly different at (A) $P < 0.01$ and (B) $P < 0.001$ (sequential Bonferroni-corrected Wilcoxon matched pairs test).

of catches in both treatments did not differ when compared with each other. Thus, 1,4-benzoquinone, neither alone nor in combination with green leaf volatiles, is attractive to *M. melolontha* males (Fig. 2.3 A).

Experiment 3: Test for attractiveness of toluquinone. Toluquinone-baited traps (21) did not catch significantly more male cockchafers than empty control traps (17). Traps baited with the green leaf volatile mixture caught significantly more beetles (132) than empty control or toluquinone-baited traps. However, when toluquinone was applied together with the green leaf volatile mixture the number of caught males (247) increased and was significantly higher than the number obtained by addition of male catches from each single treatment (Fig. 2.3 B).

2.4 Discussion

By use of gas chromatography with electroantennographic detection three potential pheromone compounds have been identified. 1,4-Benzoquinone, toluquinone and nonanal were present in dichloromethane extracts from both sexes and elicited comparable electrophysiological responses on antennae from both sexes. Preliminary results revealed that nonanal was neither attractive to male cockchafers when tested alone, nor in combination with green leaf volatiles. However, as in another melolonthine species, *Macrodactylus subspinosus* F., the closely related compounds 1-nonanol and (*E*)-2-nonenol are potent attractants (Williams *et al.* 2000), further experiments will be performed to finally elucidate the potential role of nonanal as attractant or coattractant for *M. melolontha* males.

M. melolontha females produce toluquinone and 1,4-benzoquinone, as do *M. hippocastani* females (Ruther *et al.* 2001b). In southern Germany both species are sympatric (Niklas 1960). Occasional cross-breeding has been discussed but never been proved (Niklas 1970). Swarming flight and increased mating activity can be observed in both species at dusk. In other sympatric scarab species cross-mating is avoided on the basis of divergent chemical constituents (Leal *et al.* 1996), or agonistic-antagonistic activities of enantiomeric pheromone compounds (e.g. Tumlinson *et al.* 1977; Leal 1996; Wojtasek *et al.* 1998). Toluquinone, the sex pheromone from *M. melolontha* females, is not attractive to *M. hippocastani* males (Ruther *et al.* 2001a). On the other hand, 1,4-benzoquinone is not attractive and does not enhance the attractiveness of green leaf volatiles to *M. melolontha* males (Fig. 2.3 B). However, both compounds elicit comparable electroantennographic responses in both species. Based on these results, two hypotheses come up which need further examination: (a) Not the production but the emission of sex pheromone components is species specific (1,4-benzoquinone for *M. hippocastani* and toluquinone for *M. melolontha*). (b) A species-specific blend of the identified compounds, or a still unknown chemical, allows for species-specific mate recognition and thereby species isolation.

The presented data and the results of a recent study (Reinecke *et al.* 2002) strongly suggest that the European cockchafer *M. melolontha* uses the same mate-finding strategy as recently demonstrated in the closely related forest cockchafer, *M. hippocastani* (Ruther *et al.* 2001). At dusk, males start swarming flights, whereas females remain feeding on the host trees. Green leaf volatiles are released by plants after mechanical damage (Hatanaka *et al.* 1995). In recent studies addressing the role of individual green leaf volatile compounds we demonstrated that *M. hippocastani* is only attracted by (*Z*)-3-hexen-1-ol (Ruther *et al.* 2002), whereas for *M. melolontha* other common leaf alcohols, i.e. 1-hexanol and (*E*)-2-hexen-1-ol, are attractive too (Reinecke *et al.* 2002). Thus, males are enabled to locate leaf damage sites and thereby to approach the vicinity of feeding females. In order to discriminate between leaf damage caused by feeding conspecifics and unspecific leaf damage, cockchafers use beetle-derived benzoquinones as sex pheromones. Whereas 1,4-benzoquinone has been shown to act as a sex pheromone in *M. hippocastani*, the present study demonstrates that *M. melolontha* uses toluquinone, the methyl derivative of this compound.

Toluquinone is not attractive *per se* but synergistically enhances the attractiveness of green leaf volatiles. To our knowledge this is a remarkable difference to all other so-far identified major compounds of sex pheromones in scarab beetles that are attractive *per se* (e.g. Tumlinson *et al.* 1977; Tamaki *et al.* 1985; Leal *et al.* 1992a,b, 1993a,b, 1994b; Leal 1993; Facundo *et al.* 1994; Zhang *et al.* 1994, 1996; Hallet *et al.* 1995; Kakizaki *et al.* 2000; Ruther *et al.* 2001a). Thus, plant volatiles are the primary attractants for *M. melolontha* males and function as a sexual kairomone as demonstrated before for *M. hippocastani* (Ruther *et al.* 2002).

Toluquinone was identified in extracts from both sexes. Hence, also males might emit this compound, an assumption that is supported by occasionally observed mating attempts among males. The production site of toluquinone and 1,4-benzoquinone in *Melolontha* cockchafers is still unknown. Even though toluquinone was detected in extracts from both sexes of *M. melolontha* and both sexes respond physiologically to it, we consider this compound a sex pheromone as it synergistically enhances attraction of only males towards green leaf volatiles released from (female) feeding sites. Thus, toluquinone meets the definition of a sex pheromone as given by Landolt and Phillips (1997).

An antimicrobial and antifungal activity of both toluquinone and 1,4-benzoquinone at natural concentration levels has been demonstrated against *Escherichia coli* and

Saccharomyces cerevisiae (Ruther *et al.* 2001b). Toluquinone and 1,4-benzoquinone inhibit germination of blastospores of the entomopathogenic fungi *Metarhizium anisopliae* and *Beauveria brongniartii* even though at supernatural concentration levels (Ruther *et al.* 2001b). Both 1,4-benzoquinone and toluquinone are well-known defence compounds in other arthropod taxa (e.g. Eisner 1958; Schildknecht & Holoubek 1961; Eisner & Meinwald 1966; Tschinkel 1975; Eisner *et al.* 1978; Steidle & Dettner 1993). According to the secondary function hypothesis, the role of some scarab pheromones might have evolved from other primary functions, e.g. defence (Haynes & Potter 1995; Leal 1997). Thus, toluquinone that is not attractive to *M. melolontha* males *per se*, but synergises the attractiveness of plant volatiles, may represent an evolutionary link between chemicals with merely defensive functions and sex pheromones that attract potential mates on their own.

2.5 Acknowledgements

Manfred Fröschle, Landesanstalt für Pflanzenschutz, Stuttgart, and Robert Weiland, Head of Obergrombach, gave logistic support in our field work. We thank Ute Braun and Frank Müller from our laboratory for technical assistance, our students Anke Steppuhn, Katja Hadwich and Maya Ulbricht for help in the field, and Dr Stefan Sieben, Berlin, for his contribution to the field experiments of 2001. This research was funded by the Hessian state forest administration until spring 2001 and by the Deutsche Forschungsgemeinschaft (DFG Hi 416/13–1).

2.6 References

Bauernfeind RJ, Haynes KF & Potter DA (1999). Responses of three *Cyclocephala* species to hexane extracts of *Cyclocephala lurida* sex pheromone. J Kansas Entomol Soc 72:246–247.

Eisner T (1958). Spray mechanism of the cockroach *Diploptera punctata*. Science 128:148–149.

Eisner T, Alsop K, Hichs K & Meinwald J (1978). Defensive secretions of millipeds. In: Bettini S (ed). Arthropod Venoms. Springer, Berlin, pp 41–72.

Eisner T & Meinwald J (1966). Defensive secretions of arthropods. Science 153:1341–1350.

Facundo HT, Zhang A, Robins PS, Alm SR, Linn Jr. CE, Villani MG & Roelofs WL (1994). Sex pheromone response of the Oriental beetle (Coleoptera: Scarabaeidae). Environ Entomol 23:1508–1515.

Hallett RH, Perez AL, Gries G, Gries R, Pierce Jr. HD, Yue J, Oehlschlager AC, Gonzales LM & Borden JH (1995). Aggregation pheromone of coconut rhinoceros beetle, *Oryctes rhinoceros* (L.) (Coleoptera: Scarabaeidae). J Chem Ecol 21:1549–1570.

Hatanaka A, Kajiwara T & Matsui K (1995). The biogeneration of green odour by green leaves and its physiological functions – past, present and future. Z Naturforschung 50:467–472.

Haynes KF & Potter DA (1995). Sexual response of male scarab beetle to larvae suggests a novel evolutionary origin for a pheromone. Am Entomol 41:169–175.

Hurpin B (1962). Famille des Scarabaeides. In: Balachowsky AS (ed). Entomologie appliquée à l'agriculture. Masson et Cie, Paris, pp 24–203.

Kakizaki M, Sugie H, Fukomoto T & Ino M (2000). Lure using synthetic sex pheromone, (*Z*)-7,15-hexadecadien-4-olide, of yellowish elongate chafer, *Heptophylla picea* MOTSCHULSKI (Coleoptera: Scarabaeidae). Jpn J Appl Entomol Z 44:44–46.

Keller S (1986). Biologie und Populationsdynamik, Historischer Rückblick, Kulturmassnahmen. In: Neuere Erkenntnisse über den Maikäfer. Beih Mitt Thurgau Naturforsch Ges 1:12–39.

Keller S (1993). Is there a two year development of the cockchafer *Melolontha melolontha* L.? Mitt Schweiz Entomol Ges 66:243–246.

Klein MG, Tumlinson JH, Ladd Jr. TL & Doolittle RE (1981). Japanese beetle (Coleoptera: Scarabaeidae): Response to synthetic sex attractant plus phenethyl propionate: eugenol. J Chem Ecol 7:1–7.

Krell FT (1996). The copulatory organs of the cockchafer, *Melolontha melolontha*. (Insecta: Coleoptera: Scarabaeidae). A contribution to comparative and functional anatomy of ectodermal genitalia of the coleoptera. Stuttgarter Beitr Naturkd Ser A 537:1-101.

Landolt PJ & Phillips TW (1997). Host plant influence on sex pheromone behaviour of phytophagous insects. Annu Rev Entomol 42:371–391.

Leal WS (1993). (*Z*)- and (*E*)-tetradec-7-en-2-one, a new type of sex pheromone from the oriental beetle. Naturwissenschaften 80:86–87.

Leal WS (1996). Chemical communication in scarab beetles: Reciprocal behavioral agonist–antagonist activities of chiral pheromones. P Natl Acad Sci USA 93:12112–12115.

Leal WS (1997). Evolution of sex pheromone communication in plant-feeding scarab beetles. In: Cardé RT & Minks AK (eds). Insect Pheromone Research, New Directions. Chapman & Hall, New York, pp 505–513.

Leal WS (1998). Chemical ecology of phytophagous scarab beetles. Annu Rev Entomol 43:39–61.

Leal WS, Hasegawa M & Sawada M (1992a). Identification of *Anomala schonfeldti* sex pheromone by high-resolution GC-behavior bioassay. Naturwissenschaften 79:518–519.

Leal WS, Hasegawa M, Sawada M, Ono M & Tada S (1996). Scarab beetle *Anomala albopilosa albopilosa* utilizes a more complex sex pheromone system than a similar species *A. cuprea*. J Chem Ecol 22:2001–2010.

Leal WS, Hasegawa M, Sawada M, Ono M & Ueda Y (1994b). Identification and field evaluation of *Anomala octiescostata* (Coleoptera: Scarabaeidae) sex pheromone. J Chem Ecol 20:1643–1655.

Leal WS, Matsuyama S, Kuwahara Y & Hasegawa M (1992b). An amino acid derivative as the sex pheromone of a scarab beetle. Naturwissenschaften 79:184–185.

Leal WS, Ono M, Hasegawa M & Sawada M (1994a). Kairomone from dandelion, *Taraxum officinale*, attractant for scarab beetle *Anomala octiescostata*. J Chem Ecol 20:1697–1704.

Leal WS, Sawada M & Hasegawa M (1993a). The scarab beetle *Anomala daimiana* utilizes a blend of two other *Anomala* spp. sex pheromones. Naturwissenschaften 80:181–183.

Leal WS, Sawada M, Matsuyama S, Kuwahara Y & Hasegawa M (1993b). Unusual periodicity of pheromone production in the large black chafer, *Holotrichia parallela*. J Chem Ecol 19:1381–1391.

Niklas OF (1960). Standorteinflüsse und natürliche Feinde als Begrenzungsfaktoren von. *Melolontha*-Larvenpopulationen eines Waldgebietes (Forstamt Lorsch, Hessen) (Coleoptera: Scarabaeidae). Mitt Biol Bundesanst Land & Forstw 101:5–59.

Niklas OF (1970). Die Variabilität einiger Artmerkmale von *Melolontha melolontha* (Linnaeus) und *M. hippocastani* (Fabricius) (Coleoptera: Lamellicornia: Melolonthidae). Nachrichtenbl Dtsch Pflanzenschutzdienst 22:182–189.

Reinecke A, Ruther J, Tolasch T, Francke W & Hilker M (2002). Alcoholism in cockchafers: Orientation of male *Melolontha melolontha* towards green leaf alcohols. Naturwissenschaften 89:265-269.

Ruther J, Podsiadlowski L & Hilker M (2001b). Quinones in cockchafers: Additional function of a sex attractant as an antimicrobial agent. Chemoecology 11:225–229.

Ruther J, Reinecke A & Hilker M (2002). Plant volatiles in the sexual communication of *Melolontha hippocastani*: Response towards time-dependent bouquets and novel function of (Z)-3-hexen-1-ol as a sexual kairomone. Ecol Entomol 27:76-83.

Ruther J, Reinecke A, Thiemann K, Tolasch T, Francke W & Hilker M (2000). Mate finding in the forest cockchafer, *Melolontha hippocastani*, mediated by volatiles from plants and females. Physiol Entomol 25:172–179.

Ruther J, Reinecke A, Tolasch T & Hilker M (2001a). Make love not war: A common arthropod defence compound as sex pheromone in the forest cockchafer *Melolontha hippocastani*. Oecologia 128:44–47.

Sachs L (1992). Angewandte Statistik, 7th edn. Springer Verlag, Berlin.

Schildknecht H & Holoubek K (1961). Die Bombardierkäfer und ihre Explosionschemie. Angew Chemie 73:1–7.

Steidle JLM & Dettner K (1993). Chemistry and morphology of the tergal gland of free-living adult Aleocharinae (Coleoptera: Staphylinidae) and its phylogenetic significance. Syst Entomol 18:149–168.

Tamaki Y, Sukie H & Noguchi H (1985). Methyl (*Z*)-5-tetradecenoate: Sex-attractant pheromone of the soybean beetle *Anomala rufocuprea* MOTSCHULSKY (Coleoptera: Scarabaeidae). Appl Entomol Zool 20:359–361.

Tschinkel WR (1975). A comparative study of the chemical defensive system of tenebrionid beetles: chemistry of the secretions. J Insect Physiol 21:753–783.

Tumlinson JH, Klein MG, Doolittle RE, Ladd, TL & Proveaux AT (1977). Identification of the female Japanese beetle sex pheromone: Inhibition of the male response by an enantiomer. Science 197:789–792.

Williams RN, Fickle DS, McGovern TP & Klein MG (2000). Development of an attractant for the scarab pest *Macrodactylus subspinosus* (Coleoptera: Scarabaeidae). J Econ Entomol 93:1480–1484.

Wojtasek H, Hansson BS & Leal WS (1998). Attracted or repelled? A matter of two neurons, one pheromone binding protein, and a chiral center. Biochem Biophys Res Comms 250:217–222.

Yarden G & Shani A (1994). Evidence for volatile chemical attractants in the beetle *Maladera matrida* ARGAMAN (Coleoptera: Scarabaeidae). J Chem Ecol 20:2673–2685.

Zhang A, Facundo HT, Robins PS, Linn Jr. CE, Hanula JL, Villani MG & Roelofs WL (1994). Identification and synthesis of female sex pheromone of oriental beetle, *Anomala orientalis* (Coleoptera: Scarabaeidae). J Chem Ecol 20:2415–2427.

Zhang A, Robins, PS, Leal WS, Linn Jr. CE, Villani MG & Roelofs WL (1996). Essential amino acid methyl esters: Major sex pheromone components of the cranberry white grub, *Phyllophaga anxia* (Coleoptera: Scarabaeidae). J Chem Ecol 23:231–245.

Chapter 3

Phenol - Another cockchafer attractant shared by *Melolontha hippocastani* Fabr. and *M. melolontha* L.

Joachim Ruther, Andreas Reinecke and Monika Hilker

Zeitschrift für Naturforschung (2002) **57c**: 910-913 [1]

Abstract:
The response of the two most abundant cockchafer species in central Europe, *Melolontha hippocastani* and *M. melolontha*, towards phenol, mixtures of phenol with the leaf alcohol (*Z*)-3-hexen-1-ol and the known cockchafer pheromones, 1,4-benzoquinone (*M. hippocastani*) and toluquinone (*M. melolontha*), was investigated in the field. During the swarming period at dusk, phenol attracted males of both species, and enhanced the known attraction of cockchafer males towards (*Z*)-3-hexen-1-ol. A mixture of phenol plus (*Z*)-3-hexen-1-ol was less attractive for *M. hippocastani* males than a mixture of (*Z*)-3-hexen-1-ol plus 1,4-benzoquinone, whereas phenol plus (*Z*)-3-hexen-1-ol attracted as many *M. melolontha* males as a mixture of (*Z*)-3-hexen-1-ol plus toluquinone. In both species three component mixtures containing phenol, (*Z*)-3-hexen-1-ol, and the respective benzoquinone did not capture more males than two component mixtures consisting of only (*Z*)-3-hexen-1-ol and the benzoquinone. A possible role of phenol as another cockchafer sex pheromone component is discussed.

Keywords: Scarabaeidae, sex pheromone, male attractants, *Melolontha melolontha*, *Melolontha hippocastani*.

3.1 Introduction

Cockchafers of the genus *Melolontha* (may beetles) (Coleoptera: Scarabaeidae) can be severe pests in forestry, agriculture, and horticulture. Mass breeding of the two most important species, the forest cockchafer *M. hippocastani* FABR. and the European cockchafer *M. melolontha* L., occurs currently in several parts of central Europe. The demand for environmentally sound control methods has initiated an intense research focusing on the chemical orientation of cockchafers. This research revealed an intriguing mechanism of chemically mediated mate finding for both species including orientation of males towards damage-induced plant volatiles and sex pheromones. Green leaf alcohols emitted by infested host leaves after feeding of the females attract swarming males (Ruther *et al.* 2000, 2002; Reinecke *et al.* 2002a). Beetle-derived benzoquinones (1,4-benzoquinone in *M. hippocastani* and toluquinone in *M. melolontha*) synergize the male response enabling discrimination between leaf damage caused by feeding females and unspecific leaf damage (Ruther *et al.* 2001b; Reinecke *et al.* 2002b).

In a recent study, we identified seven compounds in whole body extracts of *M. hippocastani* exhibiting physiological activity in experiments using gas chromatography with coupled electroantennographic detection (GC-EAD) (Ruther *et al.* 2001b). Among these compounds were the two mentioned benzoquinones and phenol, the first known scarab beetle sex pheromone identified in the grass grub beetle *Costelytra zealandica* (Henzell & Lowe 1970). Hence, we suggested phenol as a pheromone candidate for *Melolontha* cockchafers. We present data showing that phenol attracts males of *M. hippocastani* and *M. melolontha* in the field.

3.2 Methods and Materials

Experimental sites. Both species were investigated in heavily infested deciduous forests in southwestern Germany. Field experiments with *M. hippocastani* were performed between 24 April and 13 May 2002 in a mixed woodland area near Bürstadt, in the state of Hesse. Predominant deciduous trees were *Quercus rubra* L. and *Carpinus betulus* L. Experiments with *M. melolontha* were carried out between 25 April and 10 May 2002 in a mixed woodland near Endingen (Kaiserstuhl) in the state of Baden-Württemberg. Predominant trees in the *M. melolontha* test site were *Quercus* spp., *C. betulus*, and *Fagus sylvatica* L.

3.2.1 Funnel trap experiments

The funnel traps used in the field experiments were the same as described before (Ruther *et al.* 2000). Test chemicals dissolved in $500\,\mu L$ dichloromethane and solvent controls were applied on balls of cotton wool. Sets of four traps (3 treatments and 1 control) were randomized and arranged in a complete block design. At least $30\,min$ before the swarming period, traps of each block were placed at equivalent positions ($4-7\,m$ above the ground) of infested host trees. Captures were sexed and counted the next morning. Only captures obtained in blocks of the same design were compared statistically. Numbers of beetles trapped with each treatment were analyzed by a Friedman ANOVA and consecutive multiple Wilcoxon matched pairs tests with sequential Bonferroni correction. Statistica 4.5 scientific software was used for statistical analysis (StatSoft Inc., Hamburg, Germany).

Experiment 1. The response of *M. hippocastani* and *M. melolontha* towards the following treatments was compared: (1) 500 μL dichloromethane, (2) 5.0 mg phenol, (3) 5.0 mg (*Z*)-3-hexen-1-ol, (4) 5.0 mg each of phenol and (*Z*)-3-hexen-1-ol ($n = 44$ replicates for *M. hippocastani* and $n = 23$ for *M. melolontha*).

Experiment 2. The response of *M. hippocastani* towards the following treatments was compared: (1) 500 μL dichloromethane, (2) 5.0 mg each of phenol and (*Z*)-3-hexen-1-ol, (3) 5.0 mg each of 1,4-benzoquinone and (*Z*)-3-hexen-1-ol, (4) 5.0 mg each of phenol, 1,4-benzoquinone, and (*Z*)-3-hexen-1-ol ($n = 20$ replicates).

Experiment 3. The response of *M. hippocastani* towards the following treatments was compared: (1) 500 μL dichloromethane, (2) 5.0 mg each of phenol and (*Z*)-3-hexen-1-ol, (3) 5.0 mg each of 1,4-benzoquinone and (*Z*)-3-hexen-1-ol, (4) 2.5 mg each of phenol and 1,4-benzoquinone, plus 5.0 mg of (*Z*)-3-hexen-1-ol ($n = 20$ replicates).

Experiment 4. The response of *M. melolontha* towards the following treatments was compared: (1) 500 μL dichloromethane, (2) 5.0 mg each of phenol and (*Z*)-3-hexen-1-ol, (3) 5.0 mg each of toluquinone and (*Z*)-3-hexen-1-ol, (4) 5.0 mg each of phenol, toluquinone, and (*Z*)-3-hexen-1-ol ($n = 20$ replicates).

3.3 Results

Field data are summarized in Tab. 3.1. Like in our previous cockchafer studies (Ruther *et al.* 2000, 2001a,b, 2002; Reinecke *et al.* 2002a,b), exclusively males were captured. Traps baited with phenol, (*Z*)-3-hexen-1-ol, or a mixture of both compounds caught significantly more males of both cockchafer species than non-baited control traps (Exp. 1). For *M. hippocastani*, the attractiveness of phenol and (*Z*)-3-hexen-1-ol was statistically not distinguishable, in *M. melolontha* the leaf alcohol was more attractive than phenol. In both species a mixture of phenol + (*Z*)-3-hexen-1-ol was more attractive than the single compounds.

The mixture of 1,4-benzoquinone + (*Z*)-3-hexen-1-ol attracted significantly more *M. hippocastani* males than the mixture of phenol + (*Z*)-3-hexen-1-ol (Exp. 2 – 3). A three component mixture consisting of phenol + 1,4-benzoquinone + (*Z*)-3-hexen-1-ol at 5 mg each did not lead to a further increase of male captures. On the contrary, when 5 mg each of phenol + 1,4-benzoquinone was applied (Exp. 2), the number of captured males was decreased significantly compared to 1,4-benzoquinone + (*Z*)-3-hexen-1-ol and reached the level of phenol + (*Z*)-3-hexen-1-ol. When the dose of phenol and 1,4-benzoquinone in the three-component mixture was halved to 2.5 mg each (Exp. 3), the same tendency was visible even if the difference was no longer significant ($P = 0.053$).

In *M. melolontha* the mixture of phenol + (*Z*)-3-hexen-1-ol captured as many males as the mixture of toluquinone + (*Z*)-3-hexen-1-ol (Exp. 4). A three component lure containing phenol + toluquinone + (*Z*)-3-hexen-1-ol did not increase the number of captured males when compared to mixtures of phenol + (*Z*)-3-hexen-1-ol and toluquinone + (*Z*)-3-hexen-1-ol, respectively.

Table 3.1: Mean captures of cockchafer males in differently baited funnel traps. Numbers represent mean catches ± standard deviation. Means with different lowercase letters indicate significant differences within columns for each experiment at $P < 0.05$ (multiple Wilcoxon matched pairs test after sequential Bonferroni correction).

	Lure	*M. hippocastani*	*M. melolontha*
Exp. 1	Control	0.39 ± 0.95 a	2.96 ± 4.18 a
	5.0 mg phenol	2.52 ± 5.65 b	8.17 ± 10.67 b
	5.0 mg (Z)-3-hexen-1-ol	6.43 ± 14.34 b	11.70 ± 9.95 c
	5.0 mg each of phenol + (Z)-3-hexen-1-ol	10.66 ± 15.97 c	22.70 ± 24.11 d
	Friedman ANOVA	$\chi^2 = 46.86$; df $= 3$ $P < 0.001$; $n = 44$	$\chi^2 = 43.15$; df $= 3$ $P < 0.001$; $n = 23$
Exp. 2	Control	1.65 ± 1.87 a	
	5.0 mg each of phenol + (Z)-3-hexen-1-ol	67.90 ± 82.44 b	
	5.0 mg each of 1,4-benzoquinone + (Z)-3-hexen-1-ol	117.20 ± 80.97 c	
	5.0 mg each of 1,4-benzoquinone + phenol + (Z)-3-hexen-1-ol	65.05 ± 51.41 b	
	Friedman ANOVA	$\chi^2 = 44.34$; df $= 3$ $P < 0.001$; $n = 20$	
Exp. 3	Control	1.60 ± 1.88 a	
	5.0 mg each of phenol + (Z)-3-hexen-1-ol	40.80 ± 25.31 b	
	5.0 mg each of 1,4-benzoquinone + (Z)-3-hexen-1-ol	69.55 ± 25.44 c	
	2.5 mg each of 1,4-benzoquinone + phenol + 5 mg (Z)-3-hexen-1-ol	53.40 ± 27.03 bc	
	Friedman ANOVA	$\chi^2 = 42.66$; df $= 3$ $P < 0.001$; $n = 20$	
Exp. 4	Control		1.65 ± 1.76 a
	5.0 mg each of phenol + (Z)-3-hexen-1-ol		22.00 ± 12.49 b
	5.0 mg each of toluquinone + (Z)-3-hexen-1-ol		23.70 ± 15.56 b
	5.0 mg each of toluquinone + phenol + (Z)-3-hexen-1-ol		23.20 ± 17.88 b
	Friedman ANOVA		$\chi^2 = 35.24$; df $= 3$ $P < 0.001$; $n = 20$

3.4 Discussion

Our results demonstrate that phenol exclusively attracts males of both *M. hippocastani* and *M. melolontha*. Since phenol was identified from female whole body extracts of *M. hippocastani* (Ruther *et al.* 2001b) and *M. melolontha* (unpublished data), this compound could be considered as another sex pheromone component shared by both species. Like the sex pheromones reported before in cockchafers, i.e. 1,4-benzoquinone in *M. hippocastani* (Ruther *et al.* 2001b) and toluquinone in *M. melolontha* (Reinecke *et al.* 2002b), phenol enhances the male response towards damage-induced plant volatiles. However, phenol did not further increase male captures in both species if added to lures consisting of only (Z)-3-hexen-1-ol and the respective benzoquinone. On the contrary, in *M. hippocastani* the number of males was decreased significantly after addition of phenol (Exp. 2). A possible explanation for this phenomenon might be suboptimal ratios of the three components in the lure. It is known from Lepidoptera that species specificity in closely related sympatric species could be achieved by different blends of shared pheromone components (Cardé & Baker 1984). Since our results of Exp. 2 - 4 at present do not allow to postulate the role of phenol as another cockchafer sex pheromone unambiguously, future studies will be necessary to investigate this aspect more thoroughly. Phenol has been identified before as the sex pheromone of the grass grub beetle *C. zealandica* (Henzell & Lowe 1970) and without behaviour modifying properties in another melolonthine scarab beetle, *Holotrichia consanguinea* (Leal *et al.* 1996). Like the other sex pheromones reported in melolonthine species (Leal 1997; Ruther *et al.* 2001a), phenol is a potent antimicrobial agent. Hence, phenol as a sex pheromone in *Melolontha* cockchafers would support the secondary function hypothesis on the evolution of melolonthine sex pheromones from defensive compounds (Leal 1997).

3.5 Acknowledgements

The authors thank J. Collatz, G. Dröge, N. Hesler, S. Hülse and K. Schrank for their help during the fieldwork. Dr. H. Gossenauer-Marohn, Dr. J. Gonschorrek (Hessen-Forst, Hann.-Münden) and M. Fröschle (Landesanstalt für Pflanzenschutz Stuttgart) gave logistic support. This research was financially supported by the Hessian state forest administration and the Deutsche Forschungsgemeinschaft (Hi 416/13 - 1).

3.6 References

Cardé RT & Baker TC (1984). Sexual communication with pheromones. In: Bell WJ & Cardé RT (eds). Chemical Ecology of Insects. Chapman & Hall, New York, pp 355-386.

Henzell RF & Lowe MD (1970). Sex attractant of the grass grub beetle. Science 168:1005-1006.

Leal WS (1997). Evolution of sex pheromone communication in plant-feeding scarab beetles. In: Cardé RT & Minks AK (eds). Insect Pheromone Research, New Directions. Chapman & Hall, New York, pp 505–513.

Leal WS, Yadava CPS & Vijayvergia JN (1996). Aggregation of the scarab beetle *Holotrichia consanguinea* in response to female-released pheromone suggests secondary function hypothesis for semiochemical. J Chem Ecol 22:1557-1566.

Reinecke A, Ruther J, Tolasch T, Francke W & Hilker M (2002a). Alcoholism in cockchafers: Orientation of male *Melolontha melolontha* towards green leaf alcohols. Naturwissenschaften 89:265-269.

Reinecke A, Ruther J & Hilker M (2002b). The scent of food and defence: Green leaf volatiles and toluquinone as sex attractant mediate mate finding in the European cockchafer, *Melolontha melolontha*. Ecol Lett 5:257-263.

Ruther J, Reinecke A, Thiemann K, Tolasch T, Francke W & Hilker M (2000). Mate finding in the forest cockchafer, *Melolontha hippocastani* mediated by volatiles from plants and females. Physiol Entomol 25:172-179.

Ruther J, Podsiadlowski L & Hilker M (2001a). Quinones in cockchafers: Additional function of a sex attractant as an antimicrobial agent. Chemoecology 11:225-229.

Ruther J, Reinecke A, Tolasch T & Hilker M (2001b). Make love not war: A common arthropod defence compound as sex pheromone in the forest cockchafer, *Melolontha hippocastani*. Oecologia 128:44-47.

Ruther J, Reinecke A & Hilker M (2002). Plant volatiles in the sexual communication of *Melolontha hippocastani*: Response towards time dependent bouquets and novel function of (*Z*)-3-hexen-1-ol as a sexual kairomone. Ecol Entomol 27:76-83.

Chapter 4

Precopulatory isolation in sympatric *Melolontha* species? (Coleoptera: Scarabaeidae)

Andreas Reinecke, Joachim Ruther, and Monika Hilker

Manuscript

Abstract:
The two most abundant cockchafer species in Europe, the forest cockchafer *Melolontha hippocastani* FABR. and the European cockchafer *M. melolontha* L. tend to form calamitous mass breedings with casual reports on sympatric and simultaneous occurrence. Both species are known to use feeding-induced green leaf volatiles (GLV) as primary attractants (sexual kairomones) for mate finding. The attractiveness of GLV is enhanced by the sex pheromones 1,4-benzoquinone in *M. hippocastani* and toluquinone in *M. melolontha*. Phenol attracts males from both species. All three compounds are present in females of both species. In the present study, we performed field experiments addressing the question, whether swarming males discriminate olfactorily between conspecific and heterospecific females. Males of both species preferred females, when given the choice between females and males of the other species. However, they preferred conspecific females when females from both species were offered simultaneously. The results suggest that species-specific pheromone blends contribute to precopulatory reproductive isolation in sympatric populations of *M. melolontha* and *M. hippocastani*, but are not mutually exclusive or indispensable prerequisites for mate finding as in other insects. Finally, we show that a sexual dimorphism in flight behaviour is part of the *M. melolontha* mate finding strategy, as was already shown for *M. hippocastani*.

Keywords: Mate finding, reproductive isolation, Scarabaeidae, sex pheromone, sexual kairomone, olfactory contrast, *Melolontha*.

4.1 Introduction

The two most abundant cockchafer species in Europe, the forest cockchafer *Melolontha hippocastani* FABR. and the European cockchafer *M. melolontha* L., form calamitous mass breedings in many central parts of the continent. In some areas, populations of both species overlap (Schneider 1952; Niklas 1960; Niklas 1970; Reinecke unpublished data). A set of adult morphological traits and different larval habitats constitute species differentiation into European and forest cockchafers (Niklas 1970). *M. melolontha* females deposit eggs in the soil of open landscapes. Females of *M. hippocastani* oviposit in forest soil. Occasionally egg deposition areas may overlap (Niklas 1970). Adult ecological niches overlap in many further aspects, amongst others in the range of preferred host trees (*Quercus, Acer, Carpinus, Fagus, Castanea, Aesculus*, and others) (Niklas 1974). Individuals from both species perform a spectacular swarming flight around host treetops with a peak of mating activity at sunset. In *M. hippocastani* it was shown that males perform this flight behaviour, while females remain on the host trees where they feed or have fed (Ruther *et al.* 2001). The same sexual dimorphism in flight behaviour has been assumed for *M. melolontha* (Reinecke *et al.* 2002a), and is corroborated by data presented here. Reports of occasional cross-matings of *M. melolontha* and *M. hippocastani* exist, but have been questioned by Niklas (1970). Hence, it is unclear whether and how precopulatory reproductive isolation is achieved in *M. melolontha* and *M. hippocastani*.

Given the morphological similarity of both species and the swarming flight under low light conditions at dusk (Niklas 1970), optical cues seem unlikely to account for reproductive isolation. While intraspecific acoustic communication is common in other scarab taxa, e.g. Aphodiinae (Hirschberger 2001) and Dynastinae (Mini 1997), this form of communication is unknown in the Melolonthinae.

However, mate finding in *Melolontha* species is well known to be mediated by semiochemicals. In both *Melolontha* species, swarming males use leaf alcohols emitted by the host plant upon female feeding as primary attractants (sexual kairomones) for mate finding (Ruther *et al.* 2000, 2002a; Reinecke *et al.* 2002a,b, 2005). However, *Melolontha* females do not only attract males by green leaf volatiles (GLV) induced by their feeding activity. In addition to the attractive GLV, sex pheromones lure the males. Toluquinone has been identified as sex pheromone of *M. melolontha* (Reinecke *et al.* 2002b). This component failed to elicit behavioural response in *M. hippocastani* males (Ruther *et al.* 2001). The sex pheromone of *M. hippocastani* is 1,4-benzoquinone (Ruther *et al.* 2001.). This component does not attract *M. melolontha* males (Reinecke *et al.* 2002b). However, both toluquinone and 1,4-benzoquinone are produced by either species. In addition to the quinones, they also release phenol, which is attractive to males of both species (Ruther *et al.* 2002b).

Other sympatric scarab species of the subfamilies Melolonthinae and Rutelinae with overlapping ecological niches are known to use species-specific pheromones or pheromone blends to find conspecific mating partners and to avoid cross-attraction (Leal 1998, 1999). In order to examine whether cross mating of *M. melolontha* and *M. hippocastani* are avoided by semiochemical-based precopulatory isolation mechanisms, we conducted a series of field experiments by using the landing cage experimental set-up described by Ruther *et al.* (2000) and Reinecke *et al.* (2002b). Previous experiments with these landing cages have shown that males of *M. hippocastani* and

M. melolontha were attracted to conspecific females feeding in these cages. However, no cross-attraction experiments have been conducted so far. Thus, the following questions were addressed:

(I) Do *M. melolontha* females attract *M. hippocastani* males and vice versa?

(II) Do *Melolontha* males discriminate between females of either species?

4.2 Methods and materials

4.2.1 Sexual dimorphism in flight behaviour - Estimation of the sex ratio

On five days in May 2004, the sex ratios of flying and sitting cockchafers were estimated during 30 min of the swarming period in oak stands (*Quercus robur* L.) at Dodow, Mecklenburg-Vorpommern, Germany. For this purpose flying beetles (N = 216) were caught by use of an insect net. Sitting beetles (N = 217) were collected from the leaves. Sex ratios of flying and sitting beetles were compared by a 2∗2 chi-squared test.

4.2.2 Interspecific attraction in landing cage bioassays

The landing cage bioassay has been developed to observe the response of flying cockchafer males to conspecifics under field conditions (Ruther *et al.* 2000; Reinecke *et al.* 2002b). Wire mesh cages (160∗70∗40 mm, mesh width: 5 mm) were mounted pair-wise on twigs of infested host trees at a distance of 0.5 – 1 m. Care was taken to place the cages at equivalent positions regarding light conditions, height in the trees, and surrounding foliage. Cages were baited each with five unmated beetles, which were obtained by digging them out of the soil just prior to emergence in the beginning of the season. Beetles within the cages were allowed to feed on the foliage of the twigs. Positions of cages were randomized within each pair. Landing cages were baited 30 min before onset of the flight period. During the swarming flight, landings of beetles were counted for 30 min. Landing beetles were sexed, removed from the landing cages and kept isolated until the end of the experiment to prevent double counts. Numbers of landings on the cages were statistically compared by the Wilcoxon matched pairs test (Sachs 1992).

Exp. (A) and (C) were performed in a forest with *Quercus robur* and *Q. petraea* Liebl. close to Bellheim, Rheinland-Pfalz, Germany, between 26 and 28 April 2000. Exp. (B) and (D) were performed in a yellow plum (*Prunus domestica* L.) orchard close to Les Thons, Département des Vosges, France between 28 April and 6 May 2003.

Experiment (A): Landing cages with *M. melolontha* males and females, respectively, were exposed in a *M. hippocastani* flight area.

Experiment (B): Landing cages with *M. hippocastani* males and females, respectively, were exposed in a *M. melolontha* flight area.

Experiment (C): Landing cages with female *M. melolontha* and *M. hippocastani*, respectively, were exposed in a *M. hippocastani* flight area.

Experiment (D): Landing cages with female *M. melolontha* and *M. hippocastani*, respectively, were exposed in a *M. melolontha* flight area.

4.3 Results

4.3.1 Sexual dimorphism in flight behaviour - Estimation of the sex ratio

Among flying beetles, 95 % were males, whereas 88 % of the sitting beetles were females (df $= 1$, $\chi^2 = 299.1$, $P < 0.0001$). Thus, mainly male *M. melolontha* perform the swarming flight at dusk (Fig. 4.1).

4.3.2 Interspecific attraction in landing cage bioassays

Results of Experiments (A)–(D) are summarized in Fig. 4.2.

Experiment (A): Significantly more males of *M. hippocastani* were landing on cages baited with *M. melolontha* females (219) than on cages baited with *M. melolontha* males (66) ($n = 14$ replicates; $P = 0.0027$).

Experiment (B): A significantly higher number of male *M. melolontha* were observed to land on cages baited with *M. hippocastani* females (95) than on cages baited with *M. hippocastani* males (38) ($n = 18$ replicates; $P = 0.0277$).

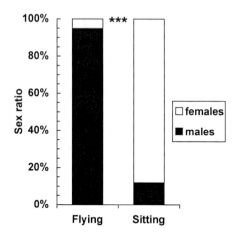

Figure 4.1: Sex ratio of flying and sitting *M. melolontha* during the swarming period at dusk. *** significance after χ^2-test with df $= 1$; $\chi^2 = 299.1$; $P < 0.001$, $n = 216$ flying and $n = 217$ persisting beeltes.

Experiment (C): *M. hippocastani* males significantly preferred females of their own species (658) to *M. melolontha* females (181) ($n = 14$ replicates; $P = 0.0035$).

Experiment (D): *M. melolontha* males significantly preferred females of their own species (111) to *M. hippocastani* females (71) ($n = 18$ replicates; $P = 0.0058$).

4.4 Discussion

The present study demonstrates that *Melolontha* males are able to discriminate between conspecific and heterospecific females. Two mechanisms of reproductive isolation mediated by pheromones are known in sympatric scarab species sharing habitats and activity patterns: (1) Female sex pheromone compounds, which attract conspecific males, may act as behavioural antagonists to males from the sympatric species (Leal 1996). (2) Reciprocal anosmia, i.e. the inability to physiologically perceive sex pheromone compounds of sympatric species, has been demonstrated as a common mechanism of reproductive isolation in scarab beetles sharing the same habitat and enantiomeric sex pheromone compounds (Leal 1999). However, both mechanisms may be excluded in cockchafers, since males show a preference for heterospecific females if exposed together with males, and male *Melolontha* antennae from both species perceive all female-derived sex attractants identified so far, i.e. 1,4-benzoquinone and

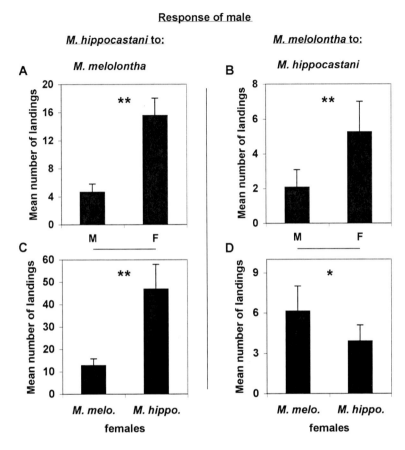

Figure 4.2: Response of male *Melolontha* cockchafers in landing cage bioassays. Mean number of males (±SE) landing during a 30 min observation period on cages baited with 5 individuals of (A) *M. melolontha* males and females, (B) *M. hippocastani* males and females, (C) and (D) female *M. melolontha* and *M. hippocastani.* Exp. (A) and exp. (C) have been performed in a *M. hippocastani* flight area; exp. (B) and exp (D) have been performed in a *M. melolontha* flight area. Significances: ∗ $P < 0.05$; ∗∗ $P < 0.01$. Statistical values are (A) $n = 14$; $P = 0.0027$; (B) $n = 14$; $P = 0.0035$; (C) $n = 18$; $P = 0.0277$; (D) $n = 18$; $P = 0.0058$; Wilcoxon matched pairs test.

toluquinone, as well as phenol, irrespective of their attractiveness in the field (Ruther *et al.* 2001; Reinecke *et al.* 2002a).

In other insects, e.g. moths, a specific pheromone blend composition may be the species' indispensable cue to elicit sexual behaviour and to find mates (Baker 1989). If this were the case in cockchafers, no preference for heterospecific females would be expected if presented together with males. However, it must be considered that feeding induced green leaf volatiles act as sexual kairomones (Ruther *et al.* 2002a). Except for the very first days of a flight season, females feed more than males (Schneider 1980). The amounts of GLV emitted by leaves females have fed upon could therefore be greater and more attractive to males if compared to the volatile amounts emitted after male feeding-damage. Additionally, each so far identified attractant enhances the attractiveness of GLV to cockchafer males (Ruther *et al.* 2001, 2002b; Reinecke *et al.* 2002b). Species-specific complex pheromone blends are therefore not an indispensable prerequisite to find mates in *M. melolontha* and *M. hippocastani*.

Nevertheless, when given the choice between con- and heterospecific females, males preferably land on cages containing females from their own species. This suggests that under the condition of overlapping populations, specific pheromone blends allow an optimized decision-making in mate searching cockchafer males. Since body extracts of both *Melolontha* species contain 1,4-benzoquinone, toluquinone, and phenol (Ruther *et al.* 2001; Reinecke *et al.* 2002b), a species specificity of a pheromone blend might be due to so far unidentified components or to quantitative differences of the known components. But our results show that differences between cockchafer pheromone blends obviously only matter, when the contrast of "right" and "wrong" female odour is present. There is only little evidence so far for the concept of olfactory contrast, which hypothesizes that a volatile component or a bouquet of volatiles only elicits a behavioural response (attraction), when perceived at the background of contrasting odour (Mumm & Hilker 2005). Whether quantitatively different blends of the identified cockchafer male attractants 1,4-benzoquinone, toluquinone, and phenol, or mixtures containing more, so far not identified compounds account for the observed phenomena, has to be answered in future investigations regarding female headspace compositions and subsequent bioassays with synthetic blends. These might lead to even more powerful lures than those described so far (Ruther *et al.* 2001, 2002b; Reinecke *et al.* 2002b).

As in *M. hippocastani*, results from counts of flying and sitting beetles during sunset show that mainly male *M. melolontha* perform the swarming flight at dusk. Thus, earlier assumptions regarding the flight behaviour in *M. melolontha* based on observations and results from *M. hippocastani* (Ruther *et al.* 2001; Reinecke *et al.* 2002a) are corroborated.

4.5 Acknowledgements

We are grateful for the support and logistical assistance during our fieldwork to Vincent Potaufeux and Jean-Luc Houot (GDEC des Vosges, Épinal), Robert Mougin, Major of Les Thons (Vosges), Manfred Kettering (Forstamt Bellheim), and Winfried Schüler (Mostobstanbau Dodow). We thank numerous student field workers for their valuable assistance in the field. This study was funded by the Deutsche Forschungsgemeinschaft (DFG, Hi 416/13-1,2) and the federal state Hesse.

4.6 References

Baker TC (1989). Sex-pheromone communication in the Lepidoptera - New research progress. Experientia 45:248-262.

Hirschberger P (2001). Stridulation in *Aphodius* dung beetles: Behavioural context and intraspecific variability of song patterns in *Aphodius ater* (Scarabaeidae). J Ins Behav 14:69-88.

Leal WS (1996). Chemical communication in scarab beetles: Reciprocal behavioral agonist-antagonist activities of chiral pheromones. P Natl Acad Sci, 93:12112-12115.

Leal WS (1998). Chemical ecology of phytophagous scarab beetles. Annu Rev Entomol 43:39-61.

Leal WS (1999). Enantiomeric anosmia in scarab beetles. J. Chem Ecol 25:1055-1066.

Mini C (1997). Scanning electron microscopic studies of the stridulatory apparatus of the coconut rhinoceros beetle *Oryctes rhinoceros* L. (Coleoptera: Scarabaeidae). Entomon 11:7-13.

Mumm R & Hilker M (2005). The significance of background odour for an egg parasitoid to detect plants with host eggs. Chem Senses 30:337-343.

Niklas OF (1960). Standorteinflüsse und natürliche Feinde als Begrenzungsfaktoren von *Melolontha* Larvenpopulationen eines Waldgebiets (Forstamt Lorsch, Hessen) (Coleoptera: Scarabaeidae). Mitt Biol Bundesanst Land- & Forstw 101:5-59.

Niklas OF (1970). Die Variabilität einiger Artmerkmale von *Melolontha melolontha* (LINNAEUS) und *M. hippocastani* (FABRICIUS) (Coleoptera: Lamellicornia: Melolonthidae). Nachrichtenbl Dtsch Pflanzenschutzdienst 22:182-189.

Niklas OF (1974). Familienreihe Lamellicornia, Blatthornkäfer. In: Schwenke W (ed). Die Forstschädlinge Europas. Parey, Hamburg und Berlin, pp 85-129.

Reinecke A, Ruther J, Tolasch T, Francke W & Hilker M (2002a). Alcoholism in cockchafers: Orientation of male *Melolontha melolontha* towards green leaf alcohols. Naturwissenschaften 89:265-269.

Reinecke A, Ruther J & Hilker M (2002b). The scent of food and defence: Green leaf volatiles and toluquinone as sex attractant mediate mate finding in the European cockchafer *Melolontha melolontha*. Ecol Letters 5:257-263.

Reinecke A, Ruther J & Hilker M (2005). Electrophysiological and behavioural response of *Melolontha melolontha* to saturated and unsaturated aliphatic alcohols. Entomol Exp Appl 115:33-40.

Ruther J, Reinecke A, Thiemann K, Tolasch T, Francke W & Hilker M (2000). Mate finding in the forest cockchafer, *Melolontha hippocastani*, mediated by volatiles from plants and females. Physiol Entomol 25:172-179.

Ruther J, Reinecke A, Tolasch T & Hilker M (2001). Make love not war: A common arthropod defence compound as sex pheromone in the forest cockchafer *Melolontha hippocastani*. Oecologia 128:44-47.

Ruther J, Reinecke A & Hilker M (2002a). Plant volatiles in the sexual communication of *Melolontha hippocastani*: Response towards time-dependent bouquets and novel function of (*Z*)-3-hexen-1-ol as a sexual kairomone. Ecol Entomol 27:76-83.

Ruther J, Reinecke A, Tolasch T & Hilker M (2002b). Phenol another cockchafer attractant shared by *Melolontha hippocastani* FABR. and *Melolontha melolontha* L. Z Naturforsch C 57:910-913.

Sachs L (1992). Angewandte Statistik, 7th edn. Springer, Berlin.

Schneider F (1952). Auftreten und Ovarialentwicklung der Maikäfer *Melolontha vulgaris* F., *M. hippocastani* F. und *M. hippocastani v. nigripes* COM. an der alpinen Verbreitungsgrenze im Hinterrheintal. Mitt Schweiz Entomol Ges 25:111-130.

Schneider P (1980). Versuche zum Fraßverhalten und zur Fraßmenge des Maikäfers, *Melolontha melolontha* L. Z Ang Entomol 90:146-161.

Chapter 5

Alcoholism in cockchafers: Orientation of male *Melolontha melolontha* towards green leaf alcohols

Andreas Reinecke, Joachim Ruther, Till Tolasch,
Wittko Francke, and Monika Hilker

Naturwissenschaften (2002) **89**: 265-269 [1]

Abstract:
Chemical orientation of the European cockchafer, *Melolontha melolontha* L., a serious pest in agriculture and horticulture, was investigated by field tests and electrophysiological experiments using plant volatiles. In total, 16 typical plant volatiles were shown to elicit electrophysiological responses in male cockchafers. Funnel trap field bioassays revealed that green leaf alcohols (i.e. (Z)-3-hexen-1-ol, (E)-2-hexen-1-ol and 1-hexanol) attracted males, whereas the corresponding aldehydes and acetates were behaviourally inactive. Furthermore, male cockchafers were attracted by volatiles from mechanically damaged leaves of *Fagus sylvatica* L., *Quercus robur* L. and *Carpinus betulus* L. However, volatiles emitted by damaged leaves of *F. sylvatica* attracted significantly more males than those from the other host plants. Odour from intact *F. sylvatica* leaves was not attractive to *M. melolontha* males. Females were not attracted by any of the tested volatile sources. The results suggest that plant volatiles play a similar role as a sexual kairomone in mate finding of *M. melolontha*, as has been shown for the forest cockchafer, *Melolontha hippocastani* FABR. Nevertheless, both species show remarkable differences in their reaction to green leaf alcohols.

Keywords: *Melolontha melolontha*, green leaf volatiles, GLV, sexual kairomone, GC-EAD.

5.1 Introduction

The European cockchafer (May beetle), *Melolontha melolontha* L. (Coleoptera: Scarabaeidae), is a polyphagous insect that shows mass outbreaks every 30 to 40 years (Keller 1986). Adults may defoliate infested host trees, but even severe losses in leaf biomass are widely compensated by secondary sprouting (Schwerdtfeger 1970). In contrast, the root-feeding larvae cause important economic losses in agriculture, horticulture and viniculture (Hurpin 1962; Keller 1986). During the past decade, massive outbreaks in central Europe have promoted the demand for sustainable control strategies which make use of kairomones and pheromones modifying the cockchafer's behaviour.

Adults of *M. melolontha* occur from mid-April until the end of May. An intensive swarming flight around host treetops can be observed at dusk of warm and dry days throughout the flight season. During these swarming flights, mating activity increases. Both males and females mate several times (Krell 1996). At the beginning of the flight season, emerging beetles perform a spectacular, optically guided flight from the areas of larval development (e.g. orchards, vineyards) to nearby forest edges (Schneider 1952).

Little is known about olfactory orientation in *M. melolontha*. In contrast, increasing knowledge has become available concerning pheromones and plant volatiles in the orientation of many other scarab beetles (Leal 1998). For example, feeding-damage-induced plant volatiles function as aggregation kairomones in the Japanese beetle, *Popillia japonica* NEWMAN (Loughrin *et al.* 1995, 1998).

Green leaf volatiles (GLV) are released by damaged plant tissue (Hatanaka *et al.* 1995). They consist of a series of saturated and monounsaturated (*Z*)-3- or (*E*)-2-configurated six-carbon aldehydes, alcohols, and esters thereof (Visser & Avé 1978; Ruther 2000). In some insects they are known to synergistically enhance the response to pheromones (Landolt & Phillips 1997). Males of the forest cockchafer, *Melolontha hippocastani* FABR. locate their mates by olfactory orientation towards the sex pheromone 1,4-benzoquinone and GLV induced by feeding females. Orientation towards both pheromone and plant kairomones allows males to discriminate sites where females feed from sites with unspecific leaf damage (Ruther *et al.* 2000, 2001). In experiments addressing the role of individual GLV, only the leaf alcohol (*Z*)-3-hexen-1-ol was attractive to *M. hippocastani*. Since (*Z*)-3-hexen-1-ol plays a key role in mate location of the forest cockchafer, it has been referred to as a sexual kairomone (Ruther *et al.* 2002).

The present study aimed to elucidate the role of plant volatiles for orientation of *M. melolontha*. The physiological response was investigated by means of gas chromatography with coupled electroantennographic detection (GC-EAD). Furthermore, we examined the behavioural response of swarming beetles to volatiles emitted by host plant leaves and synthetic GLV in the field.

5.2 Methods and materials

5.2.1 GC-EAD experiments

A series of GC-EAD runs was performed using male and female antennae. The experimental details were described by Ruther *et al.* (2000). Synthetic mixtures containing typical plant volatiles (e.g. Takabayashi *et al.* 1991; Mattiacci *et al.* 1994; Loughrin *et al.* 1997; Ruther 2000; Tab. 5.1) at concentrations of $1\,\mathrm{mg\,mL^{-1}}$ each were injected onto the column of a gas chromatograph, and EAD responses to individual compounds were recorded. Each mixture was tested on at least three different antennae of both sexes.

5.2.2 Funnel trap experiments

Experiments were performed between 29 April and 13 May 2001 at the edge of forests close to Obergrombach, Baden-Württemberg, in southern Germany.

Funnel traps used for the experiments were the same as those described by Ruther *et al.* (2000). Synthetic lures were applied 1.5 hr before the swarming period on balls of cotton wool, which were placed into round plastic boxes (ID $*$ H : 36 $*$ 29 mm) and fixed at the top of the traps. All single chemicals and mixtures were applied in 500 μL dichloromethane. Natural baits were placed into the bottles and volatiles were allowed to evaporate through the funnel. When testing the attractiveness of volatiles from intact versus damaged leaves (Exp. 1), access of trapped beetles to the leaf material was prevented by separating the plastic bottles into two chambers of equal volume with round wire meshes (5 mm mesh).

Sets of funnel traps were arranged in a complete randomized block design. Each set was placed at least 30 min before the swarming period at equivalent positions (3–5 m above ground) of an infested host tree (*Fagus sylvatica* L., *Quercus robur* L., *Quercus petraea* LIEBL., *Acer pseudoplatanus* L. or *Carpinus betulus* L.) as described by Ruther *et al.* (2000) with a minimum distance of 1.5 m between traps. Catches were sexed and counted the next morning. Numbers of beetles trapped with each treatment were analysed by a Friedman-ANOVA and sequential Bonferroni-corrected Wilcoxon matched pairs test using Statistica 4.5 scientific software (StatSoft, Hamburg, Germany).

Natural odorant sources

Experiment 1: Attractiveness of volatiles from intact vs mechanically damaged leaves. Each block (three traps) consisted of the following treatments: (1) 5 g intact leaves of *F. sylvatica*, and (2) 5 g damaged leaves of *F. sylvatica* as volatile sources, plus (3) an empty control trap. For mechanical damage, leaves were torn into several pieces, while intact leaves were picked from the twig and carefully placed into the bait chamber to avoid mechanical damage. Two hours before the swarming period, traps were baited with freshly picked or damaged leaves ($n = 23$ replications).

Experiment 2: Attractiveness of volatiles from mechanically damaged leaves of different host plants. Each block (four traps) consisted of the following treatments: damaged leaves of either (1) *F. sylvatica*, (2) *C. betulus*, or (3) *Q. robur* as volatile sources and (4) an empty control. The amount of leaves per trap and handling of leaves were the same as described in Exp. 1 ($n = 20$ replications).

Synthetic odorants

Experiment 3: Attractiveness of (Z)-3-configurated GLV. Each block (four traps) consisted of the following treatments: (1) 5 mg of (Z)-3-hexenal, (2) 5 mg of (Z)-3-hexen-

1-ol, (3) 5 mg of (Z)-3-hexenyl acetate, and (4) 500 µL dichloromethane (solvent control) (n = 19 replications).

Experiment 4: Attractiveness of (E)-2-configurated GLV. Each block (five traps) consisted of the following treatments: (1) 5 mg of (E)-2-hexenal, (2) 5 mg of (E)-2-hexen-1-ol, (3) 5 mg of (E)-2-hexenyl acetate, (4) 5 mg of a GLV mixture composed of (Z)-3-hexenyl acetate (3.2 mg), (Z)-3-hexen-1-ol (1.5 mg), (E)-2-hexenal (0.2 mg), and (Z)-3-hexenal (0.1 mg), and (5) 500 µL dichloromethane. The GLV mixture was developed to mimic the bouquet of mechanically damaged host leaves and has been shown to be highly attractive to *M. hippocastani* (Ruther *et al.* 2002) (n = 22 replications).

Experiment 5: Attractiveness of saturated GLV. Each block (five traps) consisted of the following treatments: (1) 5 mg of hexanal, (2) 5 mg of 1-hexanol, (3) 5 mg of hexyl acetate, (4) 5 mg of the GLV mixture as described in Exp. 4, and (5) 500 µL dichloromethane (n = 17 replications).

5.3 Results

5.3.1 GC-EAD experiments

The results of the GC-EAD experiments are shown in Table 5.1. In total 23 compounds were tested for electrophysiological activity using antennae from male *M. melolontha*. Sixteen compounds elicited responses up to 4 mV. With female antennae, 12 out of 18 tested compounds triggered a physiological response. A chromatogram showing the response of a male antenna to (Z)-3-hexen-1-ol and some other typical plant volatiles is shown in Fig. 5.1.

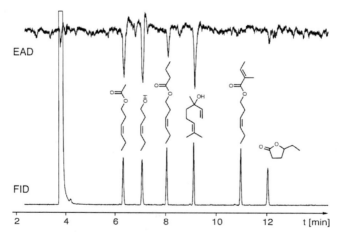

Figure 5.1: Gas chromatogram with simultaneous flame ionisation detection (FID) and electroantennographic detection (EAD) obtained by injection of 1 µL of a standard solution containing some typical plant volatiles at concentrations of 1 µg µL^{-1} each. Compounds are in order of appearance: (Z)-3-hexenyl acetate, (Z)-3-hexen-1-ol, (Z)-3-hexenyl butyrate, linalool, (Z)-3-hexenyl tiglate, γ-hexalactone

Table 5.1: GC-EAD activities of selected synthetic plant volatiles on male and female antennae of *Melolontha melolontha*. + Clear response (0.5–4 mV), (+) weak response (<0.5 mV), 0 no response, n.t. not tested.

	EAD activity	
Compound	Male	Female
(Z)-3-hexenyl acetate	+	+
(Z)-3-hexen-1-ol	+	+
(Z)-3-hexenyl butyrate	+	+
Linalool	+	+
cis-Linalool oxide	0	0
trans-Linalool oxide	0	0
(Z)-3-hexenyl tiglate	0	+
γ-hexalactone	(+)	0
(Z)-3-hexenyl benzoate	0	0
(E)-2-hexenyl acetate	(+)	n.t.
6-methyl-5-hepten-2-one	+	n.t.
(E)-2-hexen-1-ol	+	+
Benzaldehyde	(+) - +	n.t.
β-caryophyllen	0	0 - (+)
Ethyl benzoate	+	n.t.
Methyl jasmonate	0	n.t.
Hexyl acetate	+	+
1-hexanol	+	+
Hexyl hexanoate	0	0 - (+)
Methyl salicylate	+	+
Ethyl salicylate	+	+
Benzyl alcohol	+	+
2-phenylethanol	+	+

5.3.2 Funnel trap experiments

Around sunset on warm and dry days, numerous cockchafers started to fly and hover around twigs. On cold and wet days, almost no flight activity could be observed. Therefore, catch numbers varied widely from day to day depending on weather conditions. A total of 3,667 males and 126 females were caught within a period of 14 days. A preference of female cockchafers for any treatment could not be detected. Many females were observed to stay feeding on the host trees during the swarming period.

Natural odorant sources

Experiment 1: Attractiveness of volatiles from intact vs mechanically damaged leaves. Traps baited with mechanically damaged *F. sylvatica* leaves caught significantly more male beetles (144) than traps with intact leaves (40) or empty control traps (36) ($P < 0.001$; Fig. 3.2 A).

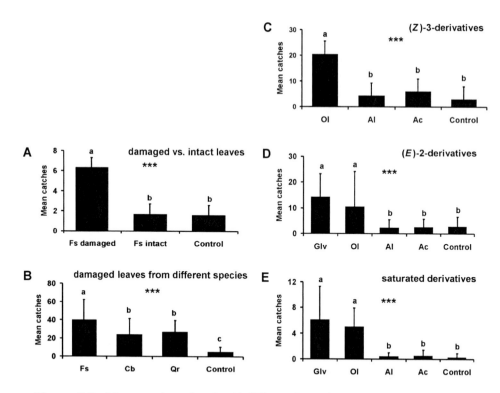

Figure 5.2: Mean number of males ($\pm$ SD) caught in funnel trap experiments with natural (A, B) and synthetic odorant sources (C–E). Blocks of traps were tested simultaneously. A–E correspond to Exp. 1–5. Fs: *F. sylvatica*, Cb: *C. betulus*, Qr: *Q. robur*, GLV: GLV mixture as described for Exp. 4 and 5, Ol: C6-alcohol, Al: C6-aldehyde, Ac: C6-acetate, Control: empty control trap in A–B and solvent control trap in C–E. $* * *$ indicates significance with (A) $\chi^2 = 22.35$, df $= 2$, $n = 23$, $P < 0.00001$; (B) $\chi^2 = 40.60$, df $= 3$, $n = 20$, $P < 0.00001$; (C) $\chi^2 = 38.48$, df $= 3$, $n = 19$, $P < 0.00001$; (D) $\chi^2 = 36.93$, df $= 4$, $n = 22$, $P < 0.00001$; (E) $\chi^2 = 41.16$, df $= 4$, $n = 17$, $P < 0.00001$ (Friedman ANOVA). Means with different letters are significantly different at (A, C) $P < 0.001$; (E) $P < 0.01$; (B) $P \leq 0.010$; (D) $P \leq 0.029$; (sequential Bonferroni-corrected Wilcoxon matched pairs test)

Experiment 2: Attractiveness of volatiles from mechanically damaged leaves of different host plants. Traps baited with damaged *F. sylvatica* leaves caught significantly more males (792) than traps baited with *C. betulus* (480) or *Q. robur* (536) ($P = 0.010$). Volatiles from mechanically damaged leaves of any tested host tree species caught significantly more male *M. melolontha* than empty control traps (103) ($P < 0.001$; Fig. 3.2 B).

Synthetic odorants

Experiment 3: Attractiveness of (Z)-3-configurated GLV. Traps baited with (Z)-3-hexen-1-ol caught significantly more male cockchafers (389) than (Z)-3-hexenal (82), (Z)-3-hexenyl acetate (114) or solvent control traps (56) ($P < 0.001$). Traps baited with (Z)-3-hexenyl acetate caught twice as many beetles as solvent control traps. Statistical analysis showed that this difference was not significant at the 0.05 level ($P = 0.069$, Fig. 3.2 C).

Experiment 4: Attractiveness of (E)-2-configurated GLV. Traps baited with (E)-2-hexen-1-ol (230) and the GLV mixture (315) caught significantly more cockchafers than (E)-2-hexenal (53), (E)-2-hexenyl acetate (67) or solvent control traps (61) ($P < 0.01$ except for (E)-2-hexen-1-ol vs control $P = 0.026$ and (E)-2-hexen-1-ol vs (E)-2-hexenal $P = 0.029$; Fig. 3.2 D).

Experiment 5: Attractiveness of saturated GLV. Traps baited with 1-hexanol (72) and the GLV mixture (87) caught significantly more cockchafers than hexanal (7), hexyl acetate (9) or solvent control traps (4) ($P < 0.01$; Fig. 3.2 E).

5.4 Discussion

The results of this study demonstrate that during the swarming period at dusk only males of the European cockchafer are attracted by volatiles emitted by mechanically damaged host plant leaves and by synthetic green leaf volatiles occurring in the bouquet of mechanically damaged host plant leaves. This suggests that in *M. melolontha* green leaf volatiles play a similar role as a sexual kairomone as in the forest cockchafer, *M. hippocastani*, enabling males to locate sites of mechanical damage caused by feeding females (Ruther *et al.* 2000, 2001, 2002). In the latter species a sexual dimorphism in the flight behaviour has been demonstrated. While male *M. hippocastani* perform the swarming flight at dusk, females remain feeding on the leaves of the host trees (Ruther *et al.* 2001). During this study, 97% of the beetles trapped were males. Therefore we assume that *M. melolontha* shows the same sexual dimorphism in flight behaviour as the closely related species.

However, there is a remarkable difference between the two *Melolontha* species regarding the role of individual GLV. *M. hippocastani* males are attracted only by (Z)-3-hexen-1-ol, while (E)-2-hexen-1-ol and 1-hexanol were behaviourally inactive (Ruther *et al.* 2002). Thus, forest cockchafer males discriminate between leaf alcohols regarding the presence, the position, and/or the configuration of a double bond in the candidate compounds. In contrast, *M. melolontha* males are attracted by any of the tested green leaf alcohols.

In both species, *M. melolontha* (the present study) and *M. hippocastani* (Ruther *et al.* 2002), males were not attracted to n-hexyl-, (E)-2-hexenyl- or (Z)-3-hexenyl acetates and aldehydes. However, at least the acetates elicited EAD signals similar to the behaviourally active alcohols.

Olfactory receptor neurons (ORNs) specific for the detection of single GLV compounds have been identified in the scarab beetle *Phyllopertha diversa* WATERHOUSE (Hansson *et al.* 1999). The response threshold to (Z)-3-hexenyl acetate was two orders of magnitude below the sensitivity of the pheromone-detecting ORNs in this species (Nikonov *et al.* 2001). The discussed ecological function of ubiquitous green leaf alcohols as sexual kairomones in *M. hippocastani* (Ruther *et al.* 2002) and *M. melolontha* may help to explain, why ORNs with a high sensitivity to GLV compounds are found in scarab beetles.

In the present study, the odorants emanating from damaged leaves of *F. sylvatica* attracted significantly more *M. melolontha* males than volatiles from leaves of the other host plants *C. betulus* and *Q. robur*. The question whether damaged *F. sylvatica* leaves release higher amounts of attractive green leaf alcohols than damaged leaves from the other host plants, or whether species specificity of the volatile pattern is responsible for their higher attractiveness, needs further investigation.

5.5 Acknowledgements

We thank Manfred Fröschle, Landesanstalt für Pflanzenschutz Stuttgart, and Robert Weiland, mayor of Obergrombach, for logistical support; Ute Braun from our laboratory for technical assistance; Maya Ulbricht, Berlin, for assistance in the field; and Dr. Stefan Sieben for his never ending enthusiasm and help in the field. Part of this work has been funded by the Deutsche Forschungsgemeinschaft (DFG, Hi 416/13-1).

5.6 References

Hansson BS, Larsson MC & Leal WS (1999). Green leaf volatile-detecting olfactory receptor neurones display very high sensitivity and specificity in a scarab beetle. Physiol Entomol 24:121–126.

Hatanaka A, Kajiwara T & Matsui K (1995). The biogeneration of green odour by green leaves and its physiological functions – past, present and future. Z Naturforsch C 50:476–472.

Hurpin B (1962). Famille des Scarabaeides. In: Balachowsky AS (ed). Entomologie appliquée à l'agriculture. Masson et Cie, Paris, pp 24–203.

Keller S (1986). Biologie und Populationsdynamik. In: Neuere Erkenntnisse über den Maikäfer. Beih Mitt Thurgau Naturforsch Ges 1:12–39.

Krell FT (1996). The copulatory organs of the cockchafer, *Melolontha melolontha* (Insecta: Coleoptera: Scarabaeidae). A contribution to comparative and functional anatomy of ectodermal genitalia of the coleoptera. Stuttgarter Beitr Naturkd Ser A 537:1-101.

Landolt PJ & Phillips TW (1997). Host plant influence on sex pheromone behavior of phytophagous insects. Annu Rev Entomol 42:371–391.

Leal WS (1998). Chemical ecology of phytophagous scarab beetles. Annu Rev Entomol 43:39–61.

Loughrin JH, Potter DA & Hamilton-Kemp TR (1995). Volatile compounds induced by herbivory act as aggregation kairomones for the Japanese beetle (*Popillia japonica* NEWMAN). J Chem Ecol 21:1457–1467.

Loughrin JH, Potter DA, Hamilton-Kemp TR & Byers ME (1997). Response of Japanese beetles (Coleoptera: Scarabaeidae) to leaf volatiles of susceptible and resistant maple species. Environ Entomol 26:334–342.

Loughrin JH, Potter DA & Hamilton-Kemp TR (1998). Attraction of Japanese beetles (Coleoptera: Scarabaeidae) to host plant volatiles in field trapping experiments. Environ Entomol 27:395–400.

Mattiacci L, Dicke M & Posthumus MA (1994). Induction of parasitoid attracting synomone in Brussels sprout plants by feeding of *Pieris brassicae* larvae: role of mechanical damage and herbivore elicitor. J Chem Ecol 20:2229–2247.

Nikonov AA, Valiyaveettil JT & Leal WS (2001). A photoaffinitylabeled green leaf volatile compound 'tricks' highly selective and sensitive insect olfactory receptor neurons. Chem Senses 26:49–54.

Ruther J (2000). Retention index database for identification of general green leaf volatiles in plants by coupled capillary gas chromatography – mass spectrometry. J Chromatogr A 890:313–319.

Ruther J, Reinecke A, Thiemann K, Tolasch T, Francke W & Hilker M (2000). Mate finding in the forest cockchafer, *Melolontha hippocastani*, mediated by volatiles from plants and females. Physiol Entomol 25:172–179.

Ruther J, Reinecke A, Tolasch T & Hilker M (2001). Make love not war: A common arthropod defense compound as sex pheromone in the forest cockchafer *Melolontha hippocastani*. Oecologia 128:44–47.

Ruther J, Reinecke A & Hilker M (2002). Plant volatiles in the sexual communication of *Melolontha hippocastani*: Response towards time-dependent bouquets and novel function of (*Z*)-3-hexen-1-ol as a sexual kairomone. Ecol Entomol 27:76-83.

Schneider F (1952). Untersuchungen über die optische Orientierung der Maikäfer (*Melolontha vulgaris* F. und *M. hippocastani* F.) sowie über die Entstehung von Schwärmbahnen und Befallskonzentrationen. Mitt Schweiz Entomol Ges 25:269–340.

Schwerdtfeger F (1970). Die Waldkrankheiten. Parey, Berlin, pp 160–163.

Takabayashi J, Dicke M & Posthumus MA (1991). Variation of predator-attracting allelochemicals emitted by herbivore-infested plants: Relative influence of plant and herbivore. Chemoecology 2:1–6.

Visser JH & Avé DA (1978). General green leaf volatiles in the olfactory orientation of the Colorado beetle, *Leptinotarsa decemlineata*. J Chem Ecol 24:538–549.

Chapter 6

Electrophysiological and behavioural responses of *Melolontha melolontha* to saturated and unsaturated aliphatic alcohols

Andreas Reinecke, Joachim Ruther, and Monika Hilker

Entomologia Experimentalis et Applicata (2005) **115**: 33-40 [1]

Abstract:

Male *Melolontha* cockchafers are known to use green leaf volatiles induced by female feeding on host plants for their mate location. Earlier studies of the response of the European cockchafer, *Melolontha melolontha* L. (Coleoptera: Scarabaeidae), to different green leaf aldehydes, alcohols, and acetates revealed that only green leaf alcohols were attractive to males in the field. Females were not attracted at all by these volatiles. Here, we present a study that aimed to elucidate the structure–activity relationships of aliphatic alcohols. Both behavioural and physiological responses were studied in male and female *M. melolontha* by field tests and electroantennography. The compounds tested were saturated aliphatic alcohols with chain lengths between five and eight carbon atoms. Furthermore, the cockchafer's response to six-carbon alcohols with (*E*)-2-, (*E*)-3-, (*Z*)-2-, (*Z*)-3-, and (*Z*)-4-configurated double bonds was tested. All compounds elicited dose-dependent responses on the antennae of both sexes. In general, males showed a stronger normalized EAG response to the stimuli than females. However, only the naturally occurring six-carbon alcohols, i.e., 1-hexanol (*E*)-2-, (*Z*)-3, and (*E*)-3-hexen-1-ol were attractive to *M. melolontha* males in the field. Females were not attracted to any of the tested compounds, confirming previous results on the olfactory orientation of *Melolontha* cockchafers.

Keywords: Sexual kairomones, green leaf volatiles, aliphatic alcohols, olfaction, electroantennogram, EAG, mate finding, Scarabaeidae.

6.1 Introduction

Interactions between insect pheromones and plant volatiles have been demonstrated in many species. Plant volatiles may trigger the release and enhance the attractiveness of sex pheromones. Thus far, pheromones have been assigned a key role in the sexual communication of scarab beetles, with plant volatiles acting as an additional cue (see reviews in: Landolt & Phillips 1997; Leal 1998; Reddy & Guerrero 2004). Physiological studies have drawn attention to the fact that scarab beetles possess olfactory receptor neurons which are highly specialized and sensitive to so-called green leaf volatiles (GLV) (Hansson *et al.* 1999; Larsson *et al.* 2001). GLV are comprised of a series of saturated and monounsaturated six-carbon aldehydes, alcohols, and esters that are formed in plant tissues following mechanical damage via the octadecanoid pathway (Gatehouse 2002). However, the ecological significance of these receptor neurons has remained unclear.

Recent studies of scarab beetles of the genus *Melolontha* (Ruther *et al.* 2000, 2002; Reinecke *et al.* 2002a) and the garden chafer *Phyllopertha horticola* (Ruther 2004) (Coleoptera: Scarabaeidae) revealed that GLV may play a key role as primary attractants in the mate location of Scarabaeidae. Males of these beetles locate females during a swarming flight by orienting towards the damage-induced GLV released by the host plants upon female feeding. This essential role of GLV during mate location might explain the presence of the above-mentioned highly GLV-sensitive olfactory receptor neurons in phytophagous scarab beetles. Plant volatiles essential for mate location have been termed sexual kairomones by Ruther *et al.* (2002). In *Melolontha* cockchafers, the male response to GLV is synergistically enhanced by beetle-derived benzoquinone derivatives enabling the males to discriminate between feeding females and unspecific leaf damage (Ruther *et al.* 2001; Reinecke *et al.* 2002a).

Studies addressing the role of individual GLV in the orientation of cockchafers revealed different results for the two species studied so far. Males of the forest cockchafer, *M. hippocastani*, responded exclusively to (*Z*)-3-hexen-1-ol (Ruther *et al.* 2002), whereas males of the European cockchafer, *M. melolontha*, were also attracted by two other naturally occurring leaf alcohols: 1-hexanol and (*E*)-2-hexen-1-ol (Reinecke *et al.* 2002b). In contrast, the corresponding aldehydes and acetates were behaviourally inactive in both species.

The present study aimed to elucidate which molecule structures of alcohols are able to elicit behavioural and physiological responses in *M. melolontha*. Both field trials and electroantennographic studies were conducted. In particular, the following questions were addressed: Is there sexual dimorphism in the dose-dependent response to leaf alcohols? How does the chain length of saturated aliphatic alcohols influence the electrophysiological and behavioural responses of *M. melolontha*? Does the presence, position, or configuration of a double bond influence the response of *M. melolontha* to mono-unsaturated six-carbon alcohols?

6.2 Methods and materials

6.2.1 Field trials

Experiments were performed between 27 April and 6 May 2003 in a yellow plum orchard at Les Thons, Département des Vosges, Eastern France.

Funnel traps were used as described by Ruther *et al.* (2000). They consisted of 2 L polyethylene wide-mouth bottles and 185 mm powder funnels (outlet diameter 40 mm) equipped with four cross-wise arranged plastic vanes standing 100 mm above the funnel. Round plastic boxes (36 mm ID * 29 mm height) were fixed on top of the vanes. Lure substances (5 mg) were dissolved in 500 μL dichloromethane and applied to balls of cotton wool which were placed in the boxes. The traps were arranged in a complete block design, with each block consisting of different treatments, a solvent control, and a reference trap baited with (*Z*)-3-hexen-1-ol, because this compound is known to attract cockchafer males effectively (Reinecke *et al.* 2002b). Traps were randomized and placed at equivalent positions into one tree per block. Height, distance from the stem, light conditions, and the development of the foliage were almost identical. The distance between traps within a block was at least 2 m. Traps were baited and placed into the trees about 1.5 hr before the flight period at sunset. Catches were sexed and counted the following morning.

Numbers of beetles caught were analysed using a nonparametric Friedman ANOVA for dependent data, followed by a Bonferroni-corrected multiple Wilcoxon matched pairs tests with trap treatment as a factor (Sachs 1992).

Influence of chain length of saturated alcohols

Experiment 1a: Test for attractiveness of 1-pentanol and 1-hexanol. Each block consisted of four traps baited with: (a)1-pentanol, (b) 1-hexanol, (c) (*Z*)-3-hexen-1-ol, and (d) solvent control ($n = 20$ replicates).

Experiment 1b: Test for attractiveness of 1-heptanol and 1-octanol. Traps within each block were baited with: (a) 1-heptanol, (b) 1-octanol, (c) (*Z*)-3-hexen-1-ol, and (d) solvent control ($n = 20$ replicates).

Influence of position and configuration of double bonds in unsaturated six-carbon alcohols

Experiment 2a: Test for attractiveness of (E)-2-hexen-1-ol and (E)-3-hexen-1-ol. Traps within each block were baited with: (a) (*E*)-2-hexen-1-ol, (b) (*E*)-3-hexen-1-ol, (c) (*Z*)-3-hexen-1-ol, and (d) solvent control ($n = 26$ replicates).

Experiment 2b: Test for attractiveness of (Z)-2-hexen-1-ol and (Z)-4-hexen-1-ol. Traps within each block were baited with: (a) (*Z*)-2-hexen-1-ol, (b) (*Z*)-4-hexen-1-ol, (c) (*Z*)-3-hexen-1-ol, and (d) solvent control ($n = 20$ replicates).

6.2.2 Electroantennographic experiments

Beetles were dug up from their hibernation sites in early spring and kept at 6°C in standard soil. They were allowed to warm up to room temperature and to feed on host plant leaves (*Carpinus betulus* L., *Quercus robur* L., *Quercus rubra* L., *Acer platanoides* L.) for a minimum of 48 hr before the experiments.

Electroantennogram measurements were made using a commercially available electroantennographic system (Syntech, Hilversum, The Netherlands) consisting of a dual electrode MTP-4 probe for antenna fixation, a CS-05 stimulus controller, and an IDAC box for data acquisition. Antennae were fanned and fixed with the tip of the distal lamella and the antennal basis between the two stainless steel electrodes using

Spectra 360 conductive gel (Parker, Orange, New Jersey). Antennae were flushed continuously with a moistened air stream (92% r.h.) via a glass tube (inner diameter 0.8 cm). Antennal preparations were inserted about 0.5 cm into the end of the tube.

Antennal responses to aliquots of standard solutions of aliphatic alcohols in dichloromethane (representing 0.1, 1, 10, and 100 μg) were recorded in concentration series with three replicates per antenna. Solvent control and reference stimuli were applied at the beginning and end of a puff series, as well as between concentration series. Responses from five antennae from different individuals were recorded per tested compound. Volatile solutions were applied on filter paper discs (5 mm in diameter) and placed into Pasteur pipettes that were connected to the stimulus controller by silicone rubber tubes. After 5 s, the solvent was blown out with a first puff. Another 5 s later, the stimulus was puffed onto the antenna by injecting the vapour phase of the Pasteur pipette 150 mm upstream from the antenna into the continuous air stream (pulse time 0.5 s, continuous flow 25 mL s^{-1} , pulse flow 21 mL s^{-1}). The minimum delay between stimulus puffs was 60 s. Ten μg of (Z)-3-hexen-ol served as a reference stimulus (100%). The following compounds were tested: 1-pentanol, 1-hexanol, 1-heptanol, 1-octanol, (E)-2-hexen-1-ol, (E)-3-hexen-1-ol, (Z)-2-hexen-1-ol, (Z)-3-hexen-1-ol, and (Z)-4-hexen-1-ol.

For data analyses, the mean solvent signal was subtracted from each mean stimulus signal. All amplitudes were normalized to the amplitude recorded in response to the reference stimulus.

Electroantennographic responses were compared statistically by two-way ANOVA followed by least significant difference tests (LSD) for the separation of means using Statistica 4.5 (StatSoft Inc., Hamburg, Germany). The antennal responses of each sex were compared regarding the factors dose and compound. Male and female responses to single compounds were compared using the factors dose and sex.

6.3 Results

6.3.1 Field trials

Trap catches varied widely depending on the weather conditions. Intensive male flight activity could be observed at sunset on warm and dry days. In contrast, only a few beetles started to hover around treetops in cold and wet weather. A total of 1836 male and 155 female *M. melolontha* were caught during the experiments. None of the tested chemicals attracted females. Thus, the numbers given below refer only to male catches.

Influence of chain length of saturated alcohols

Experiment 1a. Male catches in 1-hexanol baited traps (156) were significantly different from catches in 1-pentanol baited (23), or control traps (33). No significant difference was detected when 1-hexanol and (Z)-3-hexen-1-ol catches (211) were compared (Fig. 6.1 A).

Experiment 1b. Neither 1-heptanol- (27) nor 1-octanol baited traps (20) attracted significantly more males than solvent control traps (41), although catch numbers in (Z)-3-hexen-1-ol baited traps (228) revealed a similar flight activity as in Exp. 1a (Fig. 6.1 B).

Influence of position and configuration of double bonds in unsaturated six-carbon alcohols

Experiment 2a. (*E*)-2-hexen-1-ol attracted significantly more males (98) than the solvent control (53). However, catch numbers in (*E*)-3-hexen-1-ol (312) and (*Z*)-3-hexen-1-ol baited traps (428) exceeded this level by far. The P-level of statistical difference between these two latter catch numbers was $P = 0.058$ (Fig. 6.1 C).

Experiment 2b. Only traps baited with (*Z*)-3-hexen-1-ol (134) attracted significantly more males than controls (21). In contrast (*Z*)-2-hexen-1-ol (19) and (*Z*)-4-hexen-1-ol (32) were behaviourally inactive (Fig. 6.1 D).

6.3.2 Electroantennographic experiments

All tested compounds elicited dose-dependent responses from the antennae of both sexes. Statistical analyses (ANOVA) revealed that both the chemical structures of volatile compounds and the sex of the beetles from which the antennae were taken influenced electrophysiological responses. Antennal responses to standard stimuli varied between 1.4 and 7.0 mV for males, and 1.5 and 13.3 mV for females. Responses

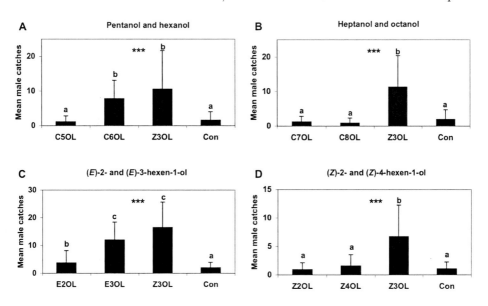

Figure 6.1: Attractiveness of aliphatic alcohols to *M. melolontha* males in funnel trap experiments (mean number of catches ±SD). Five mg of (A) 1-pentanol (C5OL) and 1-hexanol (C6OL); (B) 1-heptanol (C7OL) and 1-octanol (C8OL); (C) (*E*)-2-hexen-1-ol (E2OL) and (*E*)-3-hexen-1-ol (E3OL); and (D) (*Z*)-2-hexen-1-ol (Z2OL) and (*Z*)-4-hexen-1-ol (Z4OL) were used as bait. A solvent control (dichloromethane) (con) and a reference control with 5 mg (*Z*)-3-hexen-1-ol (Z3OL) were added to each block; ∗ ∗ ∗ indicates significance at $P < 0.001$ (Friedman ANOVA) with (A) $n = 20$, $\chi^2 = 39.419$, df = 3; (B) $n = 20$, $\chi^2 = 30.220$, df = 3; (C) $n = 26$, $\chi^2 = 56.024$, df = 3; and (D) $n = 20$, $\chi^2 = 27.976$, df = 3. Columns with different lower case letters are significantly different at $P < 0.05$ (sequential Bonferroni-corrected Wilcoxon matched pairs test).

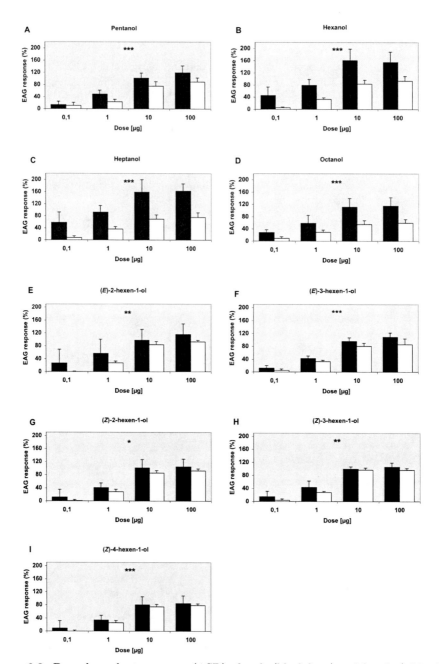

Figure 6.2: Dose-dependent responses (±SD) of male (black bars) and female (white bars) *M. melolontha* antennae to aliphatic alcohols in electroantennographic experiments. The responses for all doses were reduced by the mean response to equivalent amounts of the solvent dichloromethane. Data were subsequently normalized to the response to $10\,\mu g$ (Z)-3-hexen-1-ol as reference stimulus (= 100%). Asterisks indicate significantly different responses to the respective compound considering the factor sex (Two-way ANOVA, factors sex and dose; $n = 5$; $* P< 0.05$; $** P< 0.01$; $*** P< 0.001$). Statistical parameters of performed comparisons are listed in Table 6.1.

Table 6.1: Influence of beetle sex on electroantennographic responses –
statistical parameters after two-way ANOVA (factors: dose and sex)

Compound	Factor	MS	F	P-level
1-pentanol	dose	18819.6	95.24	< 0.0001
	sex	4684.4	23.71	< 0.00001
1-hexanol	dose	24333.8	45.98	< 0.00001
	sex	31717.8	59.93	< 0.00001
1-heptanol	dose	16967.5	30.52	< 0.00001
	sex	49692.0	89.40	< 0.00001
1-octanol	dose	10960.4	32.64	< 0.00001
	sex	16233.3	48.34	< 0.00001
(E)-2-hexen-1-ol	dose	18472.7	23.98	< 0.00001
	sex	5820.3	7.55	< 0.01
(E)-3-hexen-1-ol	dose	17465.1	157.76	< 0.00001
	sex	2002.6	18.09	< 0.001
(Z)-2-hexen-1-ol	dose	20317.6	78.81	< 0.00001
	sex	1688.3	6.55	< 0.05
(Z)-3-hexen-1-ol	dose	21078.4	171.80	< 0.00001
	sex	976.5	7.96	< 0.01
(Z)-4-hexen-1-ol	dose	14071.0	382.0	< 0.00001
	sex	593.9	16.1	< 0.001

to solvent control varied between 0.5 and 3.2 mV for males, and 0.4 and 3.2 mV for females.

Influence of beetle sex. A comparison of antennal responses to individual compounds by ANOVA (factors: sex and dose) revealed significant differences between male and female antennae for every compound used in the experiment (Fig. 6.2, Tab. 6.1). In general, male antennae were more sensitive. The statistical differences between male and female responses to individual doses are provided in Tab. 6.2 (ANOVA and subsequent LSD-test).

Influence of volatile structure. 1-Hexanol and 1-heptanol elicited a stronger response from male antennae than 1-pentanol, 1-octanol, or mono-unsaturated six-carbon alcohols, at both low and high doses (Tab. 6.3 and 6.4). Compared with 1-pentanol and 1-octanol the response to 1-heptanol was significantly stronger at every dose tested, while the response to 1-hexanol was only significantly higher than to 1-pentanol and 1-octanol at doses of 10 and 100 μg. Comparisons within the group of unsaturated alcohols revealed no significant differences, except when responses to (E)-2-hexen-1-ol and (Z)-4-hexen-1-ol at the highest dose were compared. Thus, the chain length of saturated alcohols exerted a strong influence on the male antennal response, whereas the position and configuration of the double bond in monounsaturated six-carbon alcohols only had a weak effect.

The differences in female responses to individual compounds were less pronounced than in males as far as low doses of volatiles are considered (Tab. 6.4). When 10 μg puffs of different compounds are compared, (Z)-3-hexen-1-ol elicited significantly stronger responses than any other tested compound. Responses to 1-octanol were

Table 6.2: Influence of beetle sex on electroantennographic responses to specific doses. Asterisks mark significant differences between male and female antennal responses to individual doses of the respective compounds (see Fig. 6.2). Two-way ANOVA (Tab. 6.1) and subsequent LSD-test. $*$ $P< 0.05$; $**$ $P< 0.01$; $***$ $P< 0.001$; for abbreviations see Fig. 6.1.

Dose (μg)	C5OL	C6OL	C7OL	C8OL	E2OL	E3OL	Z2OL	Z3OL	Z4OL
0.1	ns	**	**	ns	ns	ns	ns	ns	*
1	**	**	***	*	ns	ns	ns	**	*
10	**	***	***	***	ns	*	ns	ns	ns
100	**	***	***	***	ns	**	ns	ns	ns

Table 6.3: Influence of volatile structure on electroantennographic responses – statistical parameters after two-way ANOVA (factors: dose and compound).

Sex	Factor	MS	F	P-level
Males	Dose	93347.8	165,25	< 0.00001
	Compund	9607.5	17.01	< 0.00001
Fameles	Dose	67901.2	814.67	< 0.00001
	Compound	572.6	6.87	< 0.00001

Table 6.4: Influence of volatile structure on electroantennographic responses. Different lower case letters within rows indicate significantly different responses of the respective sex to alcohols at $P< 0.05$. Two-way ANOVA (Tab. 6.3) and subsequent LSD-test; for abbreviations see Fig. 6.1.

Dose (μg)	C5OL	C6OL	C7OL	C8OL	E2OL	E3OL	Z2OL	Z3OL	Z4OL
Male responses									
0.1	b	ac	a	bc	bc	b	b	b	b
1	b	ac	a	bc	bc	b	b	b	b
10	bc	a	a	b	bc	bc	bc	bc	c
100	b	a	a	b	b	bc	bc	bc	c
Female responses									
0.1	a	ab	ab	a	b	ab	ab	ab	ab
1	a	ab	b	ab	ab	ab	ab	ab	ab
10	ab	a	b	c	b	ab	b	d	ab
100	ab	a	c	d	a	abc	a	a	bc

significantly weakest at the same stimulus dose. With $100\,\mu$g puffs, maximum responses were recorded using (Z)-3-hexen-1-ol and 1-hexanol. Weakest responses were recorded with 1-heptanol, 1-octanol, and (Z)-4-hexen-1-ol as stimuli. These results indicate that chain length in saturated aliphatic alcohols, as well as the position and configuration of the double bond in mono-unsaturated six-carbon alcohols exerted an influence on female antennal responses at higher stimulus doses.

6.4 Discussion

The beetles captured in the field were almost exclusively males (92%), confirming field data from previous studies (Reinecke *et al.* 2002a,b). Female catches were randomly distributed, indicating no detectable preference for any trap treatment. This male-biased sex ratio of captured cockchafers is due to a pronounced sexual dimorphism in flight behaviour. Beetles hovering around twigs and branches in treetops during the flight period at dusk are almost exclusively males. In contrast, most individuals staying and feeding on host leaves are females (Ruther *et al.* 2001; Chap. 4). Hence, orientation towards plant-derived leaf alcohols enables mate location in *Melolontha* cockchafers.

The present study demonstrates that the male response of *M. melolontha* to alcohols is restricted to naturally occurring compounds. When considering the saturated compounds, only a chain length of six carbons (1-hexanol) was attractive. In contrast, shorter (1-pentanol) and longer (1-heptanol, 1-octanol) carbon chains turned out to be behaviourally inactive.

When considering the position and configuration of a double bond in six-carbon alcohols, *M. melolontha* males again only responded behaviourally to naturally occurring compounds. Following the mechanical damage of a plant, enzymatic degradation of C-18-fatty acids leads to the formation of the primary products hexanal and (Z)-3-hexenal via the octadecanoid pathway. The latter compound may in part rearrange to the (E)-2- and (E)-3-unsaturated derivatives (Gatehouse 2002). Reduction of the aldehydes by alcohol dehydrogenase leads to the corresponding alcohols, i.e., 1-hexanol, (Z)-3-hexen-1-ol, (E)-2-hexenol, and (E)-3-hexen-1-ol, all of which had been shown to attract male cockchafers in this study. In contrast, the two other six-carbon alcohols tested, i.e. (Z)-2- and (Z)-4-hexen-1-ol were behaviourally inactive and are to our knowledge not formed in mechanically damaged leaf tissue under natural conditions. Thus, males of *M. melolontha* are very well adapted to the naturally occurring compounds which may be used as sexual kairomones during mate finding.

In contrast to the behavioural field data, only a small influence of the volatile structure on the responsiveness was detected by the electrophysiological experiments using the electroantennography technique. All compounds tested, including the behaviourally inactive ones, elicited dose-dependent physiological responses from the antennae of both sexes. Stimuli were delivered from small filter paper disks loaded with 0.1–$100\,\mu$g of the tested compounds. The lowest tested dose was at or slightly above the EAG response threshold (Fig. 6.2). Several other lepidopteran and hymenopteran insect species showed EAG response thresholds at similar orders of magnitude, when plant volatile compounds were offered on filter paper discs, as in our study (Raguso *et al.* 1996; Park *et al.* 2001; Stelinski *et al.* 2003). In a comparable stimulation set-up, performing single sensillum recordings using males of the scarab beetle *Phyl-*

lopertha diversa WATERHOUSE, Hansson *et al.* (1999) identified receptor neurons that responded specifically to GLV, e.g., (*Z*)-3-hexen-1-ol with a stimulus load of only 10 pg. The same receptor cells responded non-specifically to other GLV at a 10.000-fold higher concentration. Olfactory receptor neurons of a similar sensitivity to GLV have been found in another scarab, *Anomala cuprea* MOTSCHULSKY (Larsson *et al.* 2001). Due to the demonstrated ecological function of leaf alcohols as sexual kairomones in *M. melolontha* and *M. hippocastani* (Reinecke *et al.* 2002b; Ruther *et al.* 2002) GLV-detecting sensilla with comparable sensitivities and specificities may also be expected to be present in these species. In this case, high stimulus doses of 0.1–100 μg would trigger many unspecific receptor neurons, explaining the low degree of selectivity in the electrophysiological responses observed in the present study.

It is generally assumed that absolute EAG responses reflect the number of activated receptor neurons and their degree of activation. In our experiments, the absolute EAG responses (mV) of female antennae were often stronger than the responses of male antennae. A sexual dimorphism of antennal morphology is common in the subfamily Melolonthinae. In *M. melolontha*, males have seven large, and females six small antennal lamellae, resulting in different distances between the tip of the distal lamella and the antennal base. Moreover, the antennal base is wider in males. Thus, male and female antennae must have different physical properties as far as electrical conductivity and resistance are concerned, which might be involved in the lower absolute EAG amplitudes recorded for males, despite the greater number of sensilla on male antennae. Furthermore, as indicated by the response range to standard stimuli (between 1.4 and 7.0 mV for males, and 1.5 and 13.3 mV for females) responses to the same stimulus varied widely between antennae. Therefore, the EAG data had to be normalized, a common procedure for maintaining comparability in EAG experiments (i.e., Anderson *et al.* 1993; Weissbecker *et al.* 1999; Jyothi *et al.* 2002; Ruther *et al.* 2002; van Tol & Visser 2002).

Our study clearly demonstrates a sexual dimorphism in the olfactory perception of green leaf volatiles by *M. melolontha*. On the basis of normalized EAG data, male antennae were more sensitive to low GLV doses, supporting the essential role that GLV play for males in the field during mate location (Fig. 6.2).

Sexual dimorphism in the perception of GLV and other plant volatiles has been observed in other insects, e.g., Braconidae, Noctuidae, Miridae, Plutellidae, and Curculionidae, but have hardly been discussed in the context of sex-specific foraging strategies or the enhancement of male responses to female sex pheromones (Evans & Allen-Williams 1992; Dickens *et al.* 1993; Chinta *et al.* 1994; Vaughn *et al.* 1996; Reddy & Guerrero 2000). It is therefore conceivable that males using GLV as sexual kairomones respond in a specific way to these compounds.

Even if the female antennae of *M. melolontha* responded in general less sensitively to GLV, when considering normalized data, the present paper and previous results (Reinecke *et al.* 2002a) suggest that both sexes are able to detect the same range of plant volatiles. The ecological context in which the females use this ability remains to be investigated, since they never responded to any of these cues during flight at dusk (Reinecke *et al.* 2002a,b). Future studies need to examine whether females use GLV for flights during other times of the day, e.g. for flights to oviposition sites. Furthermore, the possible importance of GLV, together with gustatory cues for host acceptance, will need further study.

6.5 Acknowledgements

We are grateful for support and logistical assistance during our field work from Vincent Potaufeux and Jean-Luc Houot (GDEC des Vosges, Épinal) as well as Robert Mougin, Mayor of Les Thons. We thank Ute Braun from our lab and Sandra Hülse for their help in performing the EAG experiments, and Janina Lehrke and Nadin Hermann for their valuable assistance in the field. We thank Mattias Larsson and Bernhard Weissbecker for helpful comments on an earlier draft of this manuscript. This study was funded by the Deutsche Forschungsgemeinschaft (DFG, Hi 416/13-1,2).

6.6 References

Anderson P, Hilker M, Hansson BS, Bombosch S, Klein B & Schildknecht H (1993). Oviposition deterring components in larval frass of *Spodoptera littoralis* BOISD. (Lepidoptera; Noctuidae): A behavioural and electrophysiological evaluation. J Ins Physiol 39:129–137.

Chinta S, Dickens JC & Aldrich JR (1994). Olfactory reception of potential pheromones and plant odors by tarnished plant bug, *Lygus lineolaris* (Hemiptera: Miridae). J Chem Ecol 20:3251–3267.

Dickens JC, Visser HJ & van der Pers JNC (1993). Detection and deactivation of pheromone and plant odor components by the beet armyworm, *Spodoptera exigua* HÜBNER (Lepidoptera: Noctuidae). J Ins Physiol 39:503–516.

Evans KA & Allen-Williams LJ (1992). Electroantennogram responses of the cabbage seed weevil, *Ceutorhynchus assimilis*, to oilseed rape, *Brassica napus* ssp. *oleifera*, volatiles. J Chem Ecol 18:1641–1659.

Gatehouse JA (2002). Plant resistance towards insect herbivores: A dynamic interaction. New Phytol 156:145–169.

Hansson BS, Larsson MC & Leal WS (1999). Green leaf volatile detecting olfactory receptor neurons display very high sensitivity and specificity in a scarab beetle. Physiol Entomol 24:121–126.

Jyothi KN, Prasuna AL, Sighamony S, Kumari BK, Prasad AR & Yadav JS (2002). Electroantennogram responses of *Apanteles obliquae* (Hym., Braconidae) to various infochemicals. J Appl Entomol 126:175–181.

Landolt PJ & Phillips TW (1997). Host plant influences on sex pheromone behavior of phytophagous insects. Annu Rev Entomol 42:371–391.

Larsson MC, Leal WS & Hansson BS (2001). Olfactory receptor neurons detecting plant odours and male volatiles in *Anomala cuprea* beetles (Coleoptera: Scarabaeidae). J Ins Physiol 47:1065–1076.

Leal WS (1998). Chemical ecology of phytophagous scarab beetles. Annu Rev Entomol 43:39–61.

Park KC, Zhu J, Harris J, Ochieng SA & Baker TC (2001). Electroantennogram responses of a parasitic wasp, *Microplitis croceipes*, to host-related volatile and anthropogenic compounds. Physiol Entomol 26:69–77.

Raguso RA, Light DM & Pickersky E (1996). Electroantennogram responses of *Hyles lineata* (Sphingidae: Lepidoptera) to volatile compounds from *Clarkia breweri* (Onagraceae) and other moth-pollinated flowers. J Chem Ecol 22:1735–1766.

Reddy GVP & Guerrero A (2000). Behavioral responses of the diamondback moth, *Plutella xylostella*, to green leaf volatiles of *Brassica oleracea* subsp. *capitata*. J Agri Food Chem 48:6025–6029.

Reddy GVP & Guerrero A (2004). Interactions of insect pheromones and plant semiochemicals. Trends Plant Sci 9:253–261.

Reinecke A, Ruther J & Hilker M (2002a). The scent of food and defence: Green leaf volatiles and toluquinone as sex attractant mediate mate finding in the European cockchafer *Melolontha melolontha*. Ecol Lett 5:257–263.

Reinecke A, Ruther J, Tolasch T, Francke W & Hilker M (2002b). Alcoholism in cockchafers: Orientation of male *Melolontha melolontha* towards green leaf alcohols. Naturwissenschaften 89:265–269.

Ruther J (2004). Male biassed response of garden chafer, *Phyllopertha horticola* L., to leaf alcohol and attraction of both sexes to floral plant volatiles. Chemoecology 14:187–192.

Ruther J, Reinecke A & Hilker M (2002). Plant volatiles in the sexual communication of *Melolontha hippocastani*: Response towards time-dependent bouquets and novel function of (*Z*)-3-hexen-1-ol as a sexual kairomone. Ecol Entomol 27:76–83.

Ruther J, Reinecke A, Tolasch T, Francke W & Hilker M (2000). Mate finding in the forest cockchafer, *Melolontha hippocastani*, mediated by volatiles from plants and females. Physiol Entomol 25:172–179.

Ruther J, Reinecke A, Tolasch T & Hilker M (2001). Make love not war: A common arthropod defence compound as sex pheromone in the forest cockchafer *Melolontha hippocastani*. Oecologia 128:44–47.

Sachs L (1992). Angewandte Statistik, 7th edn. Springer Verlag, Berlin.

Stelinski LL, Miller JR, Ressa NE & Gut LJ (2003). Increased EAG responses of tortricid moths after prolonged exposure to plant volatiles: Evidence for octopamine-mediated sensitisation. J Ins Physiol 49:845–856.

van Tol RWHM & Visser HJ (2002). Olfactory antennal responses of the vine weevil *Otiorhynchus sulcatus* to plant volatiles. Entomol Exp Appl 102:49–64.

Vaughn TT, Antolin MF & Bjostad LB (1996). Behavioral and physiological responses of *Diaeretiella rapae* to semiochemicals. Entomol Exp Appl 78:187–196.

Weissbecker B, van Loon JJA & Dicke M (1999). Electroantennogram responses of a predator, *Perillus bioculatus*, and its prey, *Leptinotarsa decemlineata*, to plant volatiles. J Chem Ecol 25:2313–2325.

Chapter 7

Combined effects of *Melolontha* male attractants

Andreas Reinecke, Joachim Ruther, Christoph J. Mayer,
and Monika Hilker

Journal of Applied Entomology (2006) *in press* [1]

Abstract:
Melolontha-cockchafer males search for mates using green leaf volatiles
(GLV), released by the host plants after female feeding. Thus, the feeding-
induced plant volatiles act as sexual kairomones. Males of both, *M. hippocastani* and *M. melolontha* are strongly attracted by the GLV (Z)-3-
hexen-1-ol. Sex pheromones enhance the attractiveness of (Z)-3-hexen-
1-ol and have been identified as toluquinone in *M. melolontha*, and 1,4-
benzoquinone in *M. hippocastani*. Additionally, phenol acts as male attractant in both species. Under the perspective of potential applications
we investigated by field experiments with volatile-baited traps, how to
enhance the number of captured males by the use of specific binary or
ternary blends of (Z)-3-hexen-1-ol with phenol, and toluquinone or 1,4-
benzoquinone, respectively. The presented data show that in both species
binary lures containing (Z)-3-hexen-1-ol combined with toluquinone or
1,4-benzoquinone, respectively, at a ratio of 10:1 are the most potent male
attractants.

Keywords: *Melolontha melolontha, Melolontha hippocastani*, sex pheromone, male attractant, GLV, 1,4-benzoquinone, toluquinone, phenol.

7.1 Introduction

Mass breeding of the European cockchafer, *Melolontha melolontha* L., and the forest cockchafer, *M. hippocastani* FABR., occurs currently in many parts of central Europe. Root feeding cockchafer larvae (white grubs) may cause serious losses in agriculture, horticulture, and nurseries (*M. melolontha*), or forestry (*M. hippocastani*). Adults from both species emerge at the end of April or beginning of May, depending on the local climate. They feed and mate on a variety of deciduous trees (e.g. *Quercus* spp., *Fagus sylvativa* L., *Acer* spp., *Carpinus betulus* L.). During a spectacular swarming flight at dusk, green leaf volatiles (GLV) attract males from both species to sites, where females feed (Ruther *et al.* 2000; Reinecke *et al.* 2002a), thereby acting as sexual kairomones (Ruther *et al.* 2002a). Among the GLV, (*Z*)-3-hexen-ol is among the most attractive components for males from both species (Reinecke *et al.* 2002a, 2005; Ruther *et al.* 2002a). Sex pheromones enhance the attractiveness of GLV synergistically and allow males to discriminate unspecific leaf damage from leaves with feeding conspecific females. They have been identified as 1,4-benzoquinone in *M. hippocastani* (Ruther *et al.* 2001) and toluquinone in *M. melolontha* (Reinecke *et al.* 2002b). Phenol is another beetle-derived attractant for males of both species (Ruther *et al.* 2002b).

In phytophagous scarab beetles, traps have been used as a tool in behavioural and ecological studies (Facundo *et al.* 1994, 1999), for monitoring (Alm *et al.* 1999), or control strategies (Klein & Lacey1999; Ruther & Mayer 2005). The attractiveness of the lure is, among other factors, crucial for trap efficiency. Multi-component pheromone systems are common in scarab beetles and insects from other taxa. Specific blends of the pheromone compounds, or pheromone blends and plant volatiles may enhance trap catches (e.g. Leal *et al.* 1993, 1994; Tóth *et al.* 2002; Reddy & Guerrero 2004).

In a previous study, binary mixtures of (*Z*)-3-hexen-1-ol with phenol or toluquinone, respectively, were as attractive for *M. melolontha* as a ternary mixture of all compounds in equal proportions. In contrast, the attractiveness of the lure was reduced for *M. hippocastani*, when phenol was added in equal proportions to a mixture consisting of 1,4-benzoquinone and (*Z*)-3-hexen-1-ol (Ruther *et al.* 2002b). Furthermore, an optimal ratio of 1,4-benzoquinone combined with (*Z*)-3-hexen-1-ol attractive to *M. hippocastani* was ascertained (Ruther & Hilker 2003). To provide improved tools for potential application, e.g. in cockchafer monitoring, our field experiments presented here aimed at (I) the optimization of a binary mixture of toluquinone combined with (*Z*)-3-hexen-1-ol attracting *M. melolontha* males, (II) optimization of binary mixtures of phenol combined with (*Z*)-3-hexen-1-ol attracting males from both species, and (III) optimization of ternary mixtures of (*Z*)-3-hexen-1-ol, phenol, and one of the respective quinones.

7.2 Methods and materials

7.2.1 Field sites and experimental procedures

Experimental sites. Field experiments with *M. hippocastani* were performed between 30 April and 12 May 2004 in a mixed woodland area near Bellheim, Rheinland-Pfalz in south-western Germany. Traps were placed in *Quercus rubra* L. trees. Experiments with *M. melolontha* were performed between 6 and 11 May 2004 in a yellow plum orchard close to Les Thons, département des Vosges, France (Exp. 1-2) and between 5 and 11 May 2005 in a *Q. robur* L. stand close to Dodow, Mecklenburg-Vorpommern, Germany (Exp. 3-4).

General experimental details. Behavioural responses to different blends of the beetle-derived attractants combined with the sexual kairomone (*Z*)-3-hexen-1-ol were tested in the field. Funnel traps used for the experiments were the same as described before (Ruther *et al.* 2000). They were constructed from 2 L polyethylene wide-mouthed bottles and 185 mm powder funnels (outlet diameter 40 mm) equipped with four crosswise arranged plastic vanes elevated 100 mm above the funnel. Funnel traps were baited with test volatiles, which were applied on balls of cotton wool diluted in 500 µL dichloromethane per compound and dose. They were arranged in a complete block design, with each block consisting of all treatments of the respective experiment. Traps were randomized and placed into one tree per block at equivalent positions, i.e. height, distance from the stem, light conditions, and the development of the foliage were almost identical. The distance between traps within a block was at least 2 m.

In all experiments, traps were baited and placed into the trees about 1.5 hr before the flight period at sunset. Catches were sexed and counted the following morning. Numbers of beetles caught were analysed using a nonparametric Friedman ANOVA, followed by a series of Bonferroni-corrected Wilcoxon matched pairs tests (Sachs 1992).

In a first step, optimized doses of toluquinone (*M. melolontha*) and phenol (both species) combined with 5 mg (*Z*)-3-hexen-1-ol were determined. An optimized ratio of 1,4-benzoquinone with (*Z*)-3-hexen-1-ol to attract *M. hippocastani* males has already been published (Ruther & Hilker 2003). Consecutively, we investigated whether the attractiveness of the optimized binary mixtures (1,4-benzoquinone, toluquinone, and phenol each combined with (*Z*)-3-hexen-1-ol) can be increased by adding a third component. The dose of the third compound, i.e. phenol or toluquinone in *M. melolontha*, and phenol or 1,4-benzoquinone in *M. hippocastani*, respectively, was varied in dilution steps of 1:10 between 5 and 0.005 mg. In all experiments blocks consisted of five traps. Experimental designs for *M. melolontha* and *M. hippocastani* differed in one aspect: In experiments with *M. hippocastani*, control traps were baited with either (*Z*)-3-hexen-1-ol only (Exp. 2b) or the respective optimized binary mixture of (*Z*)-3-hexen-1-ol and a beetle component (phenol or benzoquinone) (Exp. 3b and 4b). In experiments with *M. melolontha*, control traps were treated with the solvent only.

7.2.2 Binary blends

Experiment 1: Dose dependent response to toluquinone combined with 5 mg (Z)-3-hexen-1-ol (M. melolontha). Traps were baited with 5 mg (*Z*)-3-hexen-1-ol combined with 5 mg, 0.5, 0.05, and 0.005 mg toluquinone, respectively. Additionally each block contained a solvent control trap.

Experiment 2a: Dose dependent response to phenol combined with 5 mg (Z)-3-hexen-1-ol (M. melolontha). Traps were baited with 5 mg (*Z*)-3-hexen-1-ol combined with 5, 0.5, 0.05, and 0.005 mg phenol, respectively. Additionally each block contained a solvent control trap.

Experiment 2b: Dose dependent response to phenol combined with 5 mg (Z)-3-hexen-1-ol (M. hippocastani). Traps were baited with 5 mg (Z)-3-hexen-1-ol combined with 5, 0.5, 0.05, 0.005, and 0 mg phenol, respectively.

7.2.3 Ternary blends

Experiment 3a: Dose dependent response to toluquinone combined with the optimized blend of (Z)-3-hexen-1-ol and phenol (M. melolontha). Traps were baited with 5 mg (Z)-3-hexen-1-ol and 5 mg phenol combined with 5, 0.5, 0.05, and 0.005 mg toluquinone, respectively. Additionally each block contained a solvent control trap.

Experiment 3b: Dose dependent response to 1,4-benzoquinone combined with the optimized blend of (Z)-3-hexen-1-ol and phenol (M. hippocastani). Traps were baited with 5 mg (Z)-3-hexen-1-ol, and 5 mg phenol combined with 5 mg, 0.5, 0.05, 0.005, and 0 mg 1,4-benzoquinone, respectively.

Experiment 4a: Dose dependent response to phenol combined with the optimized blend of (Z)-3-hexen-1-ol and toluquinone (M. melolontha). Traps were baited with 5 mg (Z)-3-hexen-1-ol and 0.5 mg toluquinone combined with 5, 0.5, 0.05, and 0.005 mg phenol, respectively. Additionally each block contained a solvent control trap.

Experiment 4b: Dose dependent response to phenol combined with the optimized blend of (Z)-3-hexen-1-ol and 1,4-benzoquinone (M. hippocastani). Traps were baited with 5 mg (Z)-3-hexen-1-ol and 0.5 mg 1,4-benzoquinone combined with 5, 0.5, 0.05, 0.005, and 0 mg phenol, respectively.

7.3 Results

In total, 5891 *M. melolontha* and 32804 *M. hippocastani* males were caught. Total catch numbers between experiments varied widely depending on weather conditions and proceeding flight seasons. Results are summarized in Fig. 7.1. Statistical values are given in Tab. 7.1.

7.3.1 Binary blends

Experiment 1: Dose dependent response to toluquinone combined with 5 mg (Z)-3-hexen-1-ol (M. melolontha). All tested lures contained 5 mg (Z)-3-hexen-1-ol. Male catches in traps additionally baited with 0.5 mg toluquinone (463) were significantly higher than with any other treatment (Fig. 7.1 A). Traps treated with 5 mg (271), 0.05 mg (257), and 0.005 mg toluquinone (258) caught significantly different numbers of male cockchafers than the solvent control traps (38). (Number of caught beetles in brackets).

Experiment 2a: Dose dependent response to phenol combined with 5 mg (Z)-3-hexen-1-ol (M. melolontha). All tested lures contained 5 mg (Z)-3-hexen-1-ol. Traps additionally baited with 5 mg phenol (206) were significantly more attractive than any other treatment (Fig. 7.1 B). Catch numbers in traps baited with 0.5 mg (128) and 0.05 mg (86), as well as 0.05 mg and 0.005 mg phenol (72) were not significantly different from each other, whereas lures with 0.5 mg phenol attracted significantly more beetles than lures with 0.005 mg of the same compound. Catch numbers in the solvent control traps (21) were significantly lower than with any of the tested lures.

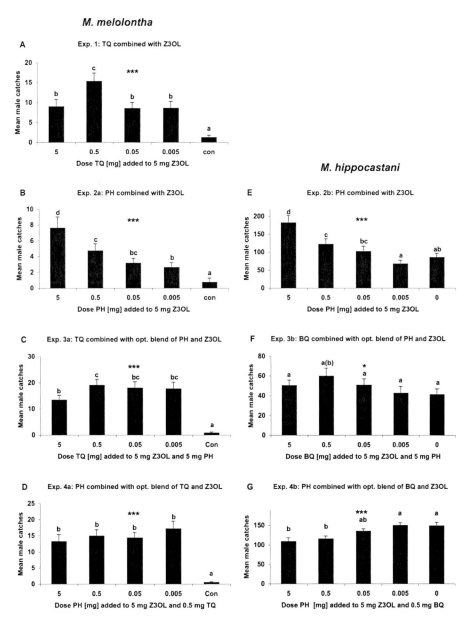

Figure 7.1: Field response of cockchafers to combined male attractants. A – D: *M. melolontha*; E – G: *M. hippocastani*. Mean number of males (±SE) in traps baited with Z3OL: (*Z*)-3-hexen-1-ol, PH: phenol, TQ: toluquinone, and BQ: 1,4-benzo-quinone; opt.: optimized. A solvent control (con) was used in bioassays with *M. melolontha*, the basic stimulus of each experiment served as control in *M. hippocastani* (0). Statistical values are given in Tab. 7.1. Asterisks indicate significant differences (*** $P < 0.001$, * $P < 0.05$, Friedman-ANOVA). Columns with different lowercase letters are significantly different at $P < 0.05$ (sequential Bonferroni-corrected Wilcoxon matched pairs test). (b) in Fig. 7.1 F indicates that the 0.5 mg column is significantly different from the control treatment, when tested doses were only compared with the control but not with each other.

Table 7.1: Field response of cockchafers to combined male attractants - Statistical values after Friedman-ANOVA (factor trap treatment); n = number of replicates.

Experiment	n	χ^2	df	P-level
1	30	44.384	4	< 0.0001
2a	27	48.202	4	< 0.0001
2b	20	53.879	4	< 0.0001
3a	32	66.056	4	< 0.0001
3b	28	10.669	4	0.0306
4a	31	64.560	4	< 0.0001
4b	21	22.291	4	0.0002

Experiment 2b: Dose dependent response to phenol combined with 5 mg (Z)-3-hexen-1-ol (M. hippocastani). All traps were baited with 5 mg (Z)-3-hexen-1-ol. The highest number of beetles was attracted when 5 mg phenol (3660) were added (Fig. 7.1 E). As in *M. melolontha*, catch numbers in traps baited with 0.5 mg (2450) and 0.05 mg (2051), as well as 0.05 mg and 0.005 mg phenol (1368) were not significantly different from each other, whereas 0.5 mg phenol attracted significantly more beetles than 0.005 mg of the same compound. The number of beetles caught in the control trap baited only with (Z)-3-hexen-1-ol (1708) did not significantly differ from catches in the 0.050 mg and 0.005 mg phenol treatments.

7.3.2 Ternary blends

Experiment 3a: Dose dependent response to toluquinone combined with the optimized blend of (Z)-3-hexen-1-ol and phenol (M. melolontha). All lures contained the optimal mixture obtained in Exp. 2a, i.e. 5 mg (Z)-3-hexen-1-ol and 5 mg phenol. Most beetles were caught in traps additionally treated with 0.5 mg toluquinone (610) (Fig. 7.1 C). However, only catches in the 5 mg treatment (432) were significantly lower, whereas numbers in the 0.05 mg (576) and 0.005 mg toluquinone treatment (567) did not differ from the 0.5 mg treatment according to statistical analysis. All tested lures attracted significantly more males than the solvent control (28).

Experiment 3b: Dose dependent response to 1,4-benzoquinone combined with the optimized blend of (Z)-3-hexen-1-ol and phenol (M. hippocastani). All traps were baited with the optimal mixture obtained in Exp. 2b, i.e. 5 mg (Z)-3-hexen-1-ol and 5 mg phenol. Comparing the 5 mg (1416), 0.5 mg (1681), 0.05 mg (1422), and 0.005 mg (1198) 1,4-benzoquinone treatments, significant differences using Bonferroni-corrected Wilcoxon tests were neither detected between these lures nor compared to the control mixture (1161) (Fig. 7.1 F). However, when the number of statistical comparisons was reduced and 1,4-benzoquinone-treated traps were compared only with the control but not with each other, the difference between the 0.5 mg 1,4-benzoquinone treatment and the control traps was significant ($P = 0.047$).

Experiment 4a: Dose dependent response to phenol combined with the optimized blend of (Z)-3-hexen-1-ol and toluquinone (M. melolontha). All lures were treated with 5 mg (Z)-3-hexen-1-ol and 0.5 mg toluquinone, the optimal mixture obtained in Exp. 1. None of the phenol treatments yielded significantly different catch numbers (412, 467, 447, 535 in order of decreasing phenol doses) (Fig. 7.1 D). All treatments caught significantly more beetles than the solvent control traps (17).

Experiment 4b: Dose dependent response to phenol combined with the optimized blend of (Z)-3-hexen-1-ol and 1,4-benzoquinone (M. hippocastani). All traps were baited with the optimized blend of 5 mg (Z)-3-hexen-1-ol and 0.5 mg 1,4-benzoquinone (Ruther & Hilker 2003). When phenol was added, the number of caught beetles decreased with higher doses (Fig. 7.1 G). Traps treated with 5 mg (2296) and 0.5 mg (2438) phenol caught significantly less beetles than the 0.005 mg (3164) and the control treatment (3139). Catch numbers with lures containing 0.05 mg phenol (2852) did not differ from any other lure.

7.4 Discussion

The overall catch numbers of 5891 *M. melolontha* and 32804 *M. hippocastani* males indicate that flight activity was widely divergent in the two species. Previous experiences with *M. melolontha* field trials indicate that under bad weather conditions and low flight activity trap catches may become erratic and thus not evaluable (Reinecke, unpublished data). Therefore, solvent control traps have been used throughout the experiments with *M. melolontha*. On the other hand, the number of equivalent trap positions available in a tree is as a rule of thumb limited to five under favourable conditions. With four different treatments and a solvent control, available trap positions were thus occupied. In forest cockchafer experiments, (Z)-3-hexen-1-ol as a well-known attractant and binary mixtures of this with a beetle pheromone component were used as control. Under dense flight activity, the latter method allows for straight comparisons between ternary and the optimized binary mixtures.

It has been shown earlier that 1,4-benzoquinone, the *M. hippocastani* sex pheromone, is most attractive to males of the species when combined at a ratio of 0.5 to 5 mg with (Z)-3-hexen-1-ol (Ruther & Hilker 2003). The presented data demonstrate that in *M. melolontha* the sex pheromone toluquinone is most attractive at the same ratio (Fig. 7.1 A).

Phenol is an attractant to males of both species, which additively enhances the attractiveness of (Z)-3-hexen-1-ol (Ruther *et al.* 2002b). Here we show that this compound is most attractive to males of both species at the highest tested dose of 5 mg per trap when combined with 5 mg (Z)-3-hexen-1-ol (Fig. 7.1 B,E).

When a ternary mixture with 5 mg each of (Z)-3-hexen-1-ol, toluquinone, and phenol was compared to binary mixtures of 5 mg (Z)-3-hexen-1-ol with either 5 mg toluquinone or phenol, all lures were equally attractive to *M. melolontha* males (Ruther *et al.* 2002b). The results presented here show that compared to a lure with 5 mg each of toluquinone and (Z)-3-hexen-1-ol, male catches were enhanced by about 70% using the optimized ratio of 0.5 mg toluquinone and 5 mg (Z)-3-hexen-1-ol. We thus infer, although no direct experimental comparison has been performed, that the optimized binary lure with toluquinone and (Z)-3-hexen-1-ol attracts more males than the lure with 5 mg each of phenol and (Z)-3-hexen-1-ol. The latter was the most attractive phenol dose, when binary phenol (Z)-3-hexen-1-ol lures were tested. Thus, among the tested binary mixtures the optimized toluquinone/(Z)-3-hexen-1-ol lure is the strongest attractant for *M. melolontha* males.

When different doses of toluquinone were tested in ternary mixtures with the optimal ratio of phenol and (Z)-3-hexen-1-ol (5 mg each), the mixture with 0.5 mg toluquinone attracted about 40% more males than the ternary mixture with 5 mg of each compound (Fig. 7.1 C). There were no statistical differences in catch numbers when the treatments with 0.5, 0.05, and 0.005 mg toluquinone, respectively, were compared. In a previous study, the ternary mixture with 5 mg of each compound attracted as many beetles as binary mixtures of 5 mg (Z)-3-hexen-1-ol with 5 mg either of phenol or toluquinone (Ruther *et al.* 2002b). Thus, we infer that addition of toluquinone at a dose of 0.5 mg to lures containing 5 mg each of (Z)-3-hexen-1-ol and phenol enhances male *M. melolontha* catches if compared to the binary lure.

Although increasing catch numbers in response to augmented phenol doses in combination with 5 mg (Z)-3-hexen-1-ol were observed in Exp. 2a, no such effect was observed once the optimized blend of 0.5 mg toluquinone and 5 mg (Z)-3-hexen-1-ol was used as basic stimulus (Fig 7.1 D). Hence, variation of phenol doses did not lead to enhanced male catch numbers, once the optimal ratio of toluquinone and (Z)-3-hexen-1-ol was applied, while the attractiveness of the optimal blend of phenol and (Z)-3-hexen-1-ol could be improved by adding a specific dose, i.e. 0.5 mg of toluquinone. Taking into account all comparisons made here, we conclude that the binary mixture of 0.5 mg toluquinone and 5 mg (Z)-3-hexen-1-ol is the most effective lure for *M. melolontha* males.

When ternary lures were tested in *M. hippocastani* flight areas, adding 0.5 mg 1,4-benzoquinone slightly enhanced the attractiveness of traps baited with the optimized blend of 5 mg each of phenol and (Z)-hexen-1-ol (Fig. 7.1 F). In an earlier study, a lure consisting of 5 mg each of 1,4-benzoquinone and (Z)-hexen-1-ol was almost twice as attractive as phenol combined with the leaf alcohol, and also twice as attractive as the ternary blend of 5 mg each of phenol, 1,4-benzoquinone, and (Z)-3-hexen-1-ol (Ruther *et al.* 2002b). Our results reported here are consistent with these earlier findings. At higher doses, i.e. 0.5 mg and 5 mg, the addition of phenol to the optimized blend of 0.5 mg 1,4-benzoquinone and 5 mg (Z)-3-hexen-1-ol deteriorates catch numbers if compared to the binary blend (Fig 7.1 G).

Thus, in both species, *M. melolontha* and *M. hippocastani*, the specific sex pheromone compounds, i.e. toluquinone and 1,4-benzoquinone, respectively, combined with (Z)-3-hexen-1-ol at a ratio of 0.5 mg : 5 mg are the strongest attractants for males. The male responses to toluquinone and 1,4-benzoquinone corroborate earlier reports that these compounds function as sex pheromones in *Melolontha* cockchafers and males from both species primarily rely on these compounds for mate location (Ruther *et al.* 2001; Reinecke *et al.* 2002b). Optimized binary blends of toluquinone or 1,4-benzoquinone with (Z)-3-hexen-1-ol attract more cockchafer males than an optimized blend of phenol and (Z)-3-hexen-1-ol. In ternary blends the addition of phenol to the optimized blends of the species' sex pheromones with (Z)-3-hexen-1-ol has either no influence (*M. melolontha*) or even deteriorates trap catches (*M. hippocastani*). However, like the two benzoquinones, phenol is present in body extracts from females of *M. hippocastani* (Ruther *et al.* 2001) and *M. melolontha* (Reinecke *et al.* 2002b), and our experiments proved that this compound in combination with (Z)-3-hexen-1-ol is an attractant for males of both species. Thus, further studies are necessary to finally understand the role of phenol in sexual communication of the two *Melolontha* species.

7.5 Acknowledgements

We are grateful for the support and logistical assistance during our field work to Vincent Potaufeux and Jean-Luc Houot (GDEC des Vosges, Épinal), Robert Mougin, Major of Les Thons (Vosges), and Winfried Schüler (Mostobstanbau Dodow). We thank Janina Lehrke, Nadin Hermann, Tanja Bloss, Marta Carboni, and Robert Koller for their valuable assistance in the field. This study was funded by the Deutsche Forschungsgemeinschaft (DFG, Hi 416/13-1,2).

7.6 References

Alm SR, Villani MG & Roelofs WL (1999). Oriental beetles (Coleoptera: Scarabaeidae): Current distribution in the United States and optimization of monitoring traps. J Econ Entomol 92:931-935.

Facundo HT, Zhang A, Robbins PS, Alm SR, Linn Jr. CE, Villani MG & Roelofs WL (1994). Sex pheromone responses of the Oriental beetle (Coleoptera: Scarabaeidae). Environ Entomol 23:1508-1515.

Facundo HT, Villani MG, Linn Jr. CE & Roelofs WL (1999). Temporal and spatial distribution of the oriental beetle (Coleoptera: Scarabaeidae) in a golf course environment. Environ Entomol 28:14-21.

Klein MG & Lacey LA (1999). An attractant trap for autodissemination of entomopathgenic fungi into populations of the Japanese beetle *Popillia japonica* (Coleoptera: Scarabaeidae). Biocontr Sci Techn 9:151-158.

Leal WS, Sawada M & Hasegawa M (1993). The scarab beetle *Anomala cuprea* utilizes the sex pheromone of *Popillia japonica* as a minor component. J Chem Ecol 19:1303-1313.

Leal WS, Kawamura F & Ono M (1994). The scarab beetle *Anomala albopilosa sakishimana* utilizes the same sex pheromone blend as a closely related and geographically isolated species *Anomala cuprea*. J Chem Ecol 20:1667-1676.

Reddy GVP & Guerrero A (2004). Interaction of insect pheromones and plant semiochemicals. Trends Plant Sci 9:253-261.

Reinecke A, Ruther J, Tolasch T, Francke W & Hilker M (2002a). Alcoholism in cockchafers: Orientation of male *Melolontha melolontha* towards green leaf alcohols. Naturwissenschaften 89:265-269.

Reinecke A, Ruther J & Hilker M (2002b). The scent of food and defence: Green leaf volatiles and toluquinone as sex attractant mediate mate finding in the European cockchafer *Melolontha melolontha*. Ecol Letters 5:257-263.

Reinecke A, Ruther J & Hilker M (2005). Electrophysiological and behavioural response of *Melolontha melolontha* to saturated and unsaturated aliphatic alcohols. Entomol Exp Appl 115:33-40.

Ruther J, Reinecke A, Thiemann K, Tolasch T, Francke W & Hilker M (2000). Mate finding in the forest cockchafer, *Melolontha hippocastani*, mediated by volatiles from plants and females. Physiol Entomol 25:172-179.

Ruther J, Reinecke A, Tolasch T & Hilker M (2001). Make love not war: A common arthropod defence compound as sex pheromone in the forest cockchafer *Melolontha hippocastani*. Oecologia 128:44-47.

Ruther J, Reinecke A & Hilker M (2002a). Plant volatiles in the sexual communication of *Melolontha hippocastani*: Response towards time-dependent bouquets and novel function of (*Z*)-3-hexen-1-ol as a sexual kairomone. Ecol Entomol 27:76-83.

Ruther J, Reinecke A, Tolasch T & Hilker M (2002b). Phenol another cockchafer attractant shared by *Melolontha hippocastani* FABR. and *Melolontha melolontha* L. Z Naturforsch C 57:910-913.

Ruther J & Hilker M (2003): Attraction of forest cockchafer *Melolontha hippocastani* to (*Z*)-3-hexen-1-ol and 1,4-benzoquinone: Application aspects. Entomol Exp Appl 107:141-147.

Ruther J & Mayer CJ (2005). Response of garden chafer, *Phyllopertha horticola*, to plant volatiles: From screening to application. Entomol Exp Appl 115:51-59.

Sachs L (1992). Angewandte Statistik, 7th edn. Springer Verlag, Berlin.

Tóth M, Furlan L, Yatsynin VG, Uiváry I, Szarukán I, Imrei Z, Subchev M, Tolasch T & Francke W (2002). Identification of sex pheromone composition of click beetle *Agriotes brevis* CANDEZE. J Chem Ecol 28:1641-1652.

Chapter 8

Do plant roots hide? Attractiveness of CO_2 to white grubs fades on the background of root exudates

Andreas Reinecke, Frank Müller, and Monika Hilker

Manuscript

Abstract:

The polyphagous larvae of the European cockchafer *Melolontha melolontha* (white grubs) feed upon plant roots, thus causing severe damages in agriculture and horticulture when calamitous mass breeding occurs. The present laboratory study investigated whether grubs orientate in the soil by chemical cues released from the roots. A new soil arena type was developed to study this question. Since former studies have shown that white grubs orientate along a CO_2 gradient, the suitability of the arena was demonstrated with CO_2 as known attractant. Grubs were significantly attracted to the test side of the arena (1.40 %) supplied with air enriched with CO_2 when compared to the control side (0.13 %) supplied with ambient air. When roots from dandelion, a highly preferred host, and roots from red clover, an accepted host, were offered, white grubs did not respond to the stimulus, even though adequate CO_2 gradients were generated by the plant roots. To study the hypothesis that root exudates interfere with the attractiveness of CO_2, roots of dandelion were extracted with an aqueous nutrient solution. When a combined stimulus of synthetic CO_2 (1.73 %) and the root extracts was offered, the white grubs were not attracted. This effect of root chemicals on the grubs´ response to CO_2 is discussed with respect to 'masking' activities of root chemicals that hide the attractiveness of CO_2.

Keywords: *Melolontha melolontha*, white grubs, soil arena, CO_2, root exudates.

8.1 Introduction

Phytophagous scarab beetle species may occur in high population densities as pest insects in agriculture, horticulture, and forestry. While adults mainly feed on above-ground foliage, the larvae may cause severe damage to the plant by feeding upon the roots. During own field experiments, dairy farmers reported losses in forage yields of up to 75 % due to larval feeding by the European cockchafer, *Melolontha melolontha*. Chemo-ecological research has largely focussed on the communication and orientation of the adults (e.g. Leal 1998 and references therein; Kim & Leal 1999; Facundo *et al.* 1999; Harari *et al.* 2000; Williams *et al.* 2000; Ruther et al. 2001, 2005; Reinecke *et al* 2002a,b, 2005; Larsson *et al.* 2003). Investigations of the soil living larvae were rather driven by the search for biological and chemical agents as well as tillage techniques to control these pests (e.g. Parasharya *et al.* 1994; Ben Yakir *et al.* 1995; Blanco & Hernandez 1995; Crutchfield *et al.* 1995; Wei 1995; Keller *et al.* 1997; Koppenhoefer & Kaya 1997; Koppenhoefer *et al.* 1999, 2000). Feeding preferences and plant-derived feeding inhibitors are well known in scarab grubs (Ene 1942; Wensler & Dudzinski 1971; Sutherland & Hillier 1974; Sutherland *et al.* 1975a,b; Crocker *et al.* 1990). For the polyphagous *M. melolontha* larvae, in particular roots of dandelion (*Taraxacum* sect. *ruderalia*) have been shown to be preferred by late larval stages. Larval feeding upon roots of this plant species leads to high weight and survival rates in *M. melolontha* (Ene 1942; Hauss 1975; Hauss & Schütte 1976; Keller 1986). However, only few investigations so far addressed the question: Do scarab larvae find roots from their hosts using chemical stimuli ?

Orientation along root exudates or root volatile gradients from intact plants has been demonstrated in numerous organisms such as roots associated bacteria, mycor-rhizal fungi, or phytoparasitic nematodes (e.g. Mateille 1994; Zheng & Sinclair 1996; Vierheilig 1998). In soil living insects, chemical cues of the rhizosphere are known to elicit behavioural responses in many species, e.g. in root feeding Diptera (*Psilae rosae, Delia radicum, Delia antiqua, Delia floralis*) and Coleoptera (*Hylastinus obscurus, Hylobius abietis, Agriotes* spp., *Hylastus nigrinus*). The stimuli belong to as different groups of chemicals as green leaf volatiles, terpenoids, isothiocyanates, aromatic compounds, and amino acids, (reviewed by Johnson & Gregory 2006). The soil-living springtail *Onichiurus armatus* utilizes specific volatiles, mostly alcohols, to locate its most preferred fungal hosts (Bengtsson *et al.* 1991). Larvae of the clover root weevil, *Sitona lepidus*, are attracted to white clover roots by an isoflavonoid, which is concentrated in N_2-fixing root nodules and potentially allows orientation not only to the roots but to specific root tissues (Johnson *et al.* 2005). Parasitoids may find their host larvae in the ground following a trail of host specific kairomones (Rogers & Potter 2002). The investigation of tritrophic sytems has recently yielded new insights into chemical communication of below-ground living organisms. The entomopathogenic nematode, *Heterorhabditis megidis*, is attracted to wine weevil larvae, *Otiorhynchus sulcatus*, by root volatiles induced upon larval feeding (van Tol *et al.* 2001). The same nematode species is lured by feeding-induced (E)-β-caryophyllene to another prey, larvae of the leaf beetle *Diabrotica virgifera virgifera* attacking maize roots (Rasmann *et al.* 2005). In another scarab, *Costelytra zealandica* (Scarabaeidae: Melolonthinae), grubs are attracted by root volatiles from *Lolium, Lotus, Trifolium,* and *Medicago*, but the active compounds were not identified (Sutherland & Hillier 1974).

CO$_2$ is an important attractant for soil living arthropods (e.g. Klingler 1958; Hasler 1986; Johnson & Gregory 2006 and references therein). *Melolontha* larvae are attracted to sources of CO$_2$, as long as a concentration gradient of at least 0.001 vol% per cm distance is present. It becomes repellent above 3 vol% (Klingler 1958; Hasler 1986). Thus, they might orient in the soil towards plant roots emitting CO$_2$ along a CO$_2$ gradient. When Hasler (1986) showed the attractiveness of CO$_2$ in *M. melolontha*, either synthetic CO$_2$ or cut root pieces were introduced into a horizontal arena. However, no natural rhizosphere was offered.

In order to study whether the orientation of *M. melolontha* larvae is guided by chemicals released from roots of living plants, a compartmented arena was designed which allowed (I) an almost undisturbed development of the plant roots even during the experiments, (II) the non-destructive yield of root exudates, (III) to monitor CO$_2$ concentrations without disturbance of the experiment, and (IV) the observation of white grubs without physical access to the plant roots, but exposed to chemical stimuli from the plant compartments. The suitability of the arena to explore white grub responses to chemical stimuli was tested using CO$_2$ as attractant. CO$_2$ gradients were monitored by using a standard gas chromatography with coupled mass spectrometry (GC-MS) method calibrated with atmospheric Argon (Ar). The response of cockchafer larvae to plant roots was investigated using roots from dandelion, *Taraxacum* sect. *ruderalia*, a highly preferred host, and roots from red clover, *Trifolium pratense* L., a species, which is still accepted (Hurpin 1962; Niklas 1974; Keller 1986). Furthermore, the response to CO$_2$ on the background of dandelion root exudates extracted from intact plant roots was explored.

8.2 Methods and materials

8.2.1 Biotest arena

Arenas (H $*$ W $*$ D $= 30 * 60 * 1$ cm) (Fig. 8.1) were constructed from Perspex. The material as well as the depth assured that the position of *M. melolontha* larvae (L3) with head widths of about 6.5 mm (Hurpin 1962) or 6.9 mm (Niklas 1974) could always be assessed. Each arena was subdivided into three compartments of equal size (W=20 cm) by nylon gauze (mesh size 50 μm) mounted on PVC frames: the outer test and control compartments, and the inner bioassay compartment. Test and control stimuli were presented in the outer compartments, while the middle compartment served as bioassay arena where the orientation of *M. melolontha* larvae was recorded. The bioassay compartment was subdivided into three fields (test, neutral, control) of equal size. The test and control fields adjoined to the respective outer compartments.

The frames separating the compartments were removable and a new gauze was taken for each bioassay. The fitting of the frames and the size of the gauze mesh ensured that plant roots could not grow through the compartment limits (Vierheilig *et al.* 1998). However, the mesh width was wide enough that both non-volatile and volatile components could enter from the outer compartments into the bioassay compartment. It was tested in Exp. 1 (see below) whether a component passing the gauze from the test compartment, is present in high concentrations in the test field and in lower

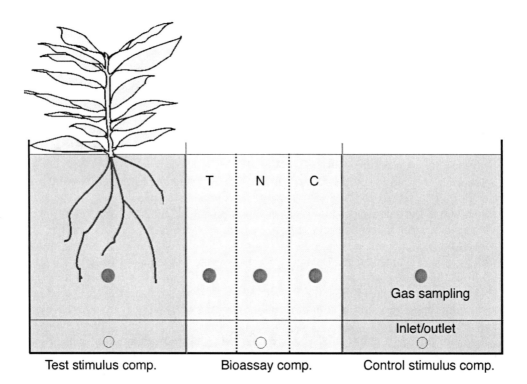

T   N   C

Gas sampling

Inlet/outlet

Test stimulus comp.   Bioassay comp.   Control stimulus comp.

Figure 8.1: Schematic drawing of the soil arena. The arena is subdivided into three compartments. Gauze (mesh size $50\,\mu$m) mounted on PVC frames delimits the middle compartment on both sides. T, N, and C label the test, neutral, and control field of the bioassay compartment. Dark dots represent gas sampling points, light dots serve as (a) gas inlets (Exp. 1 and Exp. 4), (b) drain for surplus of water (Exp. 2–4), and (c) to extract root exudates from the reservoir (Exp. 4).

ones in the control field of the bioassay compartment (i.e. test of establishment of a gradient).

The lower part of each compartment was designed as a watertight reservoir. Above the reservoirs, a lid fixed with bolts and knurled nuts allowed to open the arena on one side, without moving the compartment-limiting gauze frames. On the opposite side five small tubes (ID 2 mm) served as gas sample points. The tubes were inserted into the middle of the stimulus and the control compartment, as well as into the middle of each of the bioassay fields. Except when air samples were taken, these tubes were sealed with Teflon tape. In the lower reservoirs, tubes of the same type allowed (a) the influx of gases (Exp. 1 and Exp. 4), (b) surplus of water to drain off (Exp. 2–4), and (c) to extract root exudates from the reservoir (Exp. 4).

Flexible black overhangs protected arena bodies from light. Except for the stimulus compartments containing plants, which were tested at different ages in Exp. 2 and Exp. 3, the substrate was removed, and arenas were thoroughly cleaned between experiments.

8.2.2 Supply of arena with air

In Exp. 1 and 4, test and control stimulus compartments were supplied with air of definite CO_2 concentrations via separate tube systems. To control the flow velocity, pinchcocks (type Hoffmann) and bubble counters filled with paraffin oil were inserted into the tube systems in front of each compartments air inlet. The flow was set to $11.0\,mL\,min^{-1}$ per compartment, thus exchanging the total gas phase of a compartment every 25 min. Greenhouse air with regular CO_2 concentration was charcoal-filtered and added to the control stimulus compartments as described above. Charcoal-filtered air enriched with pressurized CO_2 (Lange Gas Service GmbH, Berlin) was added to the test stimulus compartments. The CO_2 concentration in this airflow was set to 2.1 vol% by the use of flowmeters (see below). The arenas were equilibrated with the test and control air 24 hr before the experiments started by placing the grubs into the neutral field of the middle bioassay compartment.

8.2.3 Substrate

Vermiculite, a pure mineral (Kramer Progetha, Düsseldorf; www.vermiculite.org), was chosen as substrate inside the arena. Prior to use, it was heated for a minimum of 2 hr at $180°C$. Except for stimulus compartments with living plants, arenas were filled with vermiculite before the experiments. Per compartment, 800 mL of the substrate were evenly compressed to final volume of about 540 mL, thereby reaching a density that has previously proven to be suitable for white grub locomotion as well as plant root growth.

8.2.4 Animals

White grubs were dug up at different locations in Germany and kept individually in potting soil at $12–14°C$. They were fed carrot pieces *ad libitum*. Seven days before use in the experiments, L3 were allowed to adapt to ambient temperature. *Melolonhta* larvae cease feeding activity before transforming into the pre-pupae (Niklas 1974). Therefore, only individuals, which manifestly fed the preceding days, were starved for 48 hr at room temperature and subsequently used in the experiments. White grubs were used once and had no prior experience in the experimental setup.

8.2.5 Plants

All plants were grown from seeds in a greenhouse under summer light conditions (L : D = 16 : 8 hr) using vermiculite as substrate. Five weeks after sowings, ten seedlings were planted into each test stimulus compartment and grown hydroponically with a nutrient solution modified from Arnon & Hoagland (1940) (Macronutrients: K_2NO_3 $1.02\,g\,L^{-1}$; $Ca(NO_3)_2 * 4H_2O$ $0.708\,g\,L^{-1}$; $NH_4H_2PO_4$ $0.23\,g\,L^{-1}$; $MgSO_4 * 7H_2O$ $0.49\,g\,L^{-1}$. Micronutrients: H_3BO_3 $2.860\,mg\,L^{-1}$; $MnSO_4 * 1H_2O$ $1.545\,mg\,L^{-1}$; $CuSO_4$ $* 5H_2O$ $0.080\,mg\,L^{-1}$; $ZnSO_4 * 7H_2O$ $0.220\,mg\,L^{-1}$; $Na_2MoO_4 * 2H_2O$ $0.121\,mg\,L^{-1}$; $NaFeEDTA * 2H_2O$ 12.791 $mg\,L^{-1}$). Dandelion, *Taraxacum* sectio *ruderalia*, seeds were harvested from different plants from a site heavily infested with *M. melolontha* white grubs (Kiechlinsbergen, Baden-Württemberg, Germany). Red clover, *Trifolium pratense*, seeds were purchased from a local garden supplier.

8.2.6 Determination of CO_2 concentrations

Air samples were taken from the arena by using gas-tight 500 μL syringes (SGE GmbH, Darmstadt). The syringe was inserted into the gas sample tubes of the arena described above. For calibration, synthetic CO_2 was mixed with air from which CO_2 was removed (see below). In any case, prior to collection of air samples of which CO_2 should be measured, the syringe was rinsed twice with the gas sample to collect before intake of 400–500 μL sample volume and immediate sealing of the tip with a rubber septum. The sample volume was always reduced to 100 μL prior to injection.

Carbon dioxide contents were determined by gas chromatography with coupled mass spectrometry (GC-MS), and subsequent comparison of base peak areas of CO_2 and argon. The latter served as internal standard constantly present in atmospheric air and soil cores (Sheppard & Lloyd 2002; Park *et al.* 2004). A GC (Fisons 8060, Thermoquest) was equipped with a 25 m $*$ 0.32 mm ID CP-Porabond-Q fused silica column, film thickness 5 μm (Chrompack, Varian Deutschland GmbH, Darmstadt, Germany). Helium was used as carrier gas (head pressure 10 kPa). Volumes of 100 μL were injected in splitless mode at an injector temperature of 240°C and a constant oven temperature of 40°C. The column effluent was ionized by electron impact ionization (EI) at 70 eV. The MS (Fisons MD800 quadrupole mass spectrometer) was set to selective ion mode (SIR), recording masses of 40 (Ar) and 44 (CO_2). Peak areas of the respective ions were determined using the MS software (Masslab, Thermoquest, Egelsbach, Germany).

In order to establish a calibration curve for quantification of CO_2 in the air samples taken from the arena, defined CO_2 concentrations were obtained as follows: Greenhouse air was pumped through flasks (500 mL) containing saturated aqueous $Ba(OH)2$ solutions, where natural CO_2 precipitated as $BaCO_3$. The different concentrations of CO_2 were obtained by mixing this air of which CO_2 has been removed and synthetic CO_2 at definite flows (mL min^{-1}). The gas flow was controlled by Supelco flowmeters (N032-41 and N062-01 with carboloy or glass floats, Sigma-Aldrich, Taufkirchen, Germany). CO_2/Ar peak area ratios of nine defined CO_2 concentrations ranging from 0.00 vol% to 6.07 vol% were measured. Five samples were taken from each concentration. CO_2 concentrations of gas samples taken from the arena were calculated by comparing sample CO_2/Ar peak area ratios with those obtained from the calibration. CO_2 concentration gradients between sample points were estimated assuming roughly linear gradient characteristics (distance = 40 cm between sample points in the outer compartments; distance = 13.3 cm between sample points in the test and control fields of the bioassay compartment).

8.2.7 General experimental procedures

Bioassays were conducted in a greenhouse between 10 am and 6 pm. To control effects of external factors, stimulus and control compartments were alternately oriented between consecutive arenas. At the beginning of each experiment, white grubs were placed in the middle of the neutral field at $\frac{1}{3}$ of the height of the substrate. For this purpose, the front lid of the respective arena was carefully moved upwards, creating as little disturbance as possible. A groove was pressed into the vermiculite, a white grub placed into it, and the front lid was closed. At the start, and every 15 min during a period of 6 hr, white grubs positions were recorded. The frequencies of white grub encounters in test and control fields were summarized. Numbers of encounters in test and control fields were statistically compared using the Wilcoxon matched pairs test. Each experiment was performed with $n = 20$ replicates.

For each bioassay starting 1 hr after white grubs were introduced into the bioassay arena, gas samples were taken from the stimulus and the control compartments of a subset of arenas to determine the CO_2 gradients in the arenas. CO_2 concentrations were statistically compared by the Wilcoxon matched-pairs test for dependent samples ($n = 10$ replicates in Exp. 1, 2, and 3; $n = 6$ replicates in Exp. 4)

8.2.8 Experiments

Experiment 1: Test of arena functionality with attractive CO_2. Arenas were filled with vermiculite as described above. The substrate was humidified with *aqua dest.* (200 mL per compartment). Under these conditions, the previously assessed volume of the gas phase was 280 mL per compartment. Charcoal filtered greenhouse air enriched with pressurized CO_2 to a content of about 2.1 vol% was pumped into the stimulus compartment, whereas filtered greenhouse air without enrichment served as control stimulus. The test and control stimulus compartments were equilibrated for a period of 24 hr prior to the experiments.

In addition to the measurements of CO_2 concentration in the test and control stimulus compartments (see above), CO_2 samples were taken from each field (test, neutral, control) in the middle bioassay compartment to check whether a significant CO_2 gradient builds up from the test to the control side within the white grub bioassay area. Concentration values were statistically compared by a Friedman-ANOVA with subsequent Bonferroni-corrected multiple Wilcoxon matched pairs tests.

Experiment 2: Do dandelion roots attract white grubs? T. sect. *ruderalia* plants were tested for attractiveness during week 9, 11, and 13 after sowings. During the last 24 hr before the experiments, nutrient solution (300 mL) was stepwise added to the control stimulus compartment (without plant) and the test stimulus compartment (with plant). To compensate for the plants' water and nutrient consumption, small amounts of further nutrient solution were repeatedly added to the stimulus compartment within the 24-hr-period prior to the experiment. This procedure visibly ensured that water from the test stimulus and control stimulus compartment entered the bioassay compartment equally from both sides. A small visible water film was kept at equal height at the bottom of the test and control compartment. Thus, the vermiculite was equally moistened in the test and control compartment. When the experiment started, the bioassay compartment was also evenly moisturized.

Experiment 3: Do red clover roots attract white grubs? Experimental procedures were identical to those of Exp. 2. Red clover, *T. pratense*, plants were used instead of dandelion.

Experiment 4: Does CO_2 attract white grubs in the presence of dandelion root exudates? In this experiment, the test compartment was free of plant roots, but provided with 300 mL exudates from dandelion roots in nutrient solution. Furthermore, CO_2 enriched charcoal-filtered air (2.1 %) was added to the test compartment (flow rate: 11 mL min^{-1}) as in Exp. 1. The control compartment was provided with charcoal-filtered air at the same flow rate (no CO_2 enrichment) and 300 mL nutrient solution only. Thus, the bioassay compartment was equally moisturized at the beginning of the experiments.

Root exudates were collected from 11-week-old dandelion plants grown in arenas reserved for this purpose. The plants were watered with the nutrient solution described above and used during their entire development. At the bottom of the compartments with plants, liquid from the substrate was collected in Erlenmeyer flasks via the tubes inserted into the compartments reservoir. The liquid contains the nutrients and, most probably, water-soluble components released from the roots. The liquid was collected over a period of about 4 hr until a volume of about 300 mL was obtained. The extracts were immediately ice cooled and protected from light until use.

Prior to the experiment, the gas phase of the control and test arena compartments was equilibrated for 24 hr with filtered greenhouse air and CO_2 enriched air, respectively. Additionally, root exudates were equilibrated overnight with CO_2 enriched air.

8.3 Results

Throughout all experiments, CO_2 concentrations measured in test and control compartments were significantly different from each other (Tab. 8.1). Gradients between stimulus and control compartment in Exp. 2 and Exp. 3 were at least $0.013\,\mathrm{vol\%\,cm^{-1}}$, thus one order of magnitude higher than the minimum required to elicit locomotion along a CO_2 gradient in white grubs ($0.001\,\mathrm{vol\%\,cm^{-1}}$) (Hasler 1986).

Experiment 1: Demonstration of arena functionality with attractive CO_2. *M. melolontha* larvae were significantly more often recorded in the test fields (221), close to the CO_2 source, than in the control fields (95). CO_2 concentrations in test ($1.40\,\mathrm{vol\%} \pm 0.238$) and control compartments ($0.13\,\mathrm{vol\%} \pm 0.011$) (Tab. 8.1) allowed to estimate gradients between measuring points of about $0.032\,\mathrm{vol\%\,cm^{-1}}$. Significantly different CO_2 concentrations between test, neutral, and control fields were ascertained in the bioassay compartment (Fig. 8.2). Corresponding gradients between sample points in the test and control fields in the bioassay compartment would range about $0.049\,\mathrm{vol\%\,cm^{-1}}$.

Experiment 2: Do dandelion roots attract white grubs? During the observation period, white grubs were not attracted to dandelion roots (Tab. 8.1). Total encounter numbers in test vs. control fields were 118 : 79, 47 : 66, and 87 : 52 during week 9, 11, and 13 after sowings, respectively. CO_2 concentrations in the stimulus compartments were lower during week 9, and higher during weeks 11 and 13 than in Exp. 1 (Tab. 8.1). The minimal gradients between sampling points in the stimulus and control compartments were estimated to range about $0.013\,\mathrm{vol\%/cm}$ during week 9.

Experiment 3: Do red clover roots attract white grubs? During the observation period, white grubs were not attracted to red clover roots (Tab. 8.1). Total encounter numbers in test vs. control fields were 67 : 81, 113 : 71, 114 : 42 during week 9, 11, and 13, respectively. CO_2 concentrations in the stimulus compartments were below those of Exp. 1 during week 9, and about the same as in Exp. 1 during week 11 and 13 (Tab. 8.1) with a minimum gradient between sampling points of about $0.013\,\mathrm{vol\%\,cm^{-1}}$ during week 9.

Experiment 4: Does CO_2 *attract white grubs in the presence of dandelion root exudates?* When CO_2–enriched air was used as test stimulus on the background of

Table 8.1: Response of *M. melolontha* larvae to synthetic CO_2, stimuli from plant roots of different age, and root exudates combined with CO_2. White grubs positions were noted every 15 min during 6 hr after the experiment started and summarized per individual for test and control fields. Numbers given in the white grub orientation section are means of encounters in the respective fields ($\pm$ SE) of the bioassay compartment and were statistically compared using the Wilcoxon matched pairs test ($n = 20$ replicates). CO_2 concentrations were measured in the stimulus and the control compartments of a subset of arenas parallel to the experiments. Numbers in the table are means $\pm$ SE [vol%] (Exp. 1–3: $n = 10$ replicates; Exp. 4: $n = 6$ replicates). See Fig. 8.1 for arena details.

Stimulus	Plant age	White grub orientation			CO_2 concentration		
		Test field	Control field	P	Stimulus compartment	Control compartment	P
Exp. 1: CO_2		11.1 ± 1.90	4.8 ± 1.43	*	1.40 ± 0.238	0.13 ± 0.012	**
Exp. 2: Dandelion roots	9	5.9 ± 1.79	4.0 ± 1.49	n.s.	0.71 ± 0.107	0.20 ± 0.025	**
	11	2.4 ± 1.09	3.3 ± 1.31	n.s.	2.79 ± 0.348	0.30 ± 0.043	**
	13	4.4 ± 1.67	2.6 ± 1.12	n.s.	2.35 ± 0.234	0.30 ± 0.022	**
Exp. 3: Red clover roots	9	3.4 ± 1.32	4.1 ± 1.62	n.s.	0.64 ± 0.044	0.13 ± 0.007	**
	11	5.7 ± 1.58	3.6 ± 1.45	n.s.	1.56 ± 0.103	0.36 ± 0.170	**
	13	5.7 ± 1.61	2.1 ± 1.13	n.s.	1.37 ± 0.106	0.24 ± 0.016	**
Exp. 4: Dandelion root exudates & CO_2	11	8.9 ± 2.07	6.1 ± 1.63	n.s.	1.73 ± 0.323	0.11 ± 0.026	*

Exp. 1: CO_2 gradient in the bioassay compartment

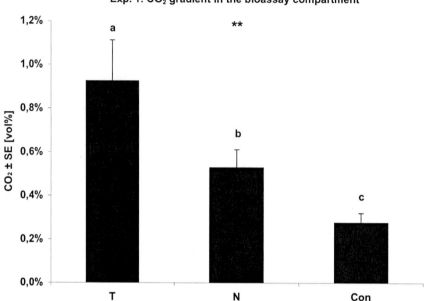

Figure 8.2: CO_2 gradient established in the bioassay compartments during Exp. 1. Cleaned air enriched with synthetic CO_2 and cleaned air were introduced into the stimulus and the control compartments of the arenas, respectively. Concentration values are given in Tab 8.1. Friedman-ANOVA ($n = 10$; df$=2$; $\chi^2 = 11.40$; $P = 0.0034$); different lower case letters indicate significant differences between concentrations after Bonferroni-corrected multiple Wilcoxon matched-pairs tests with $P < 0.05$.

root exudates, the number of total white grub encounters in the test field (177) was not significantly different from the number of encounters in the control field (122). CO_2 concentrations were about 1.73 vol% in the stimulus and 0.11 vol% in the control compartment.

8.4 Discussion

Our results show that *M. melolontha* larvae are able to orient to sources of CO_2, however, this orientation vanishes when plant roots or root exudates are present. The suitability of the arena designed here for host location experiments with *M. melolontha* larvae was especially shown by the data obtained in Exp 1. The CO_2 concentration data of this experiment demonstrate the build-up of a gradient between stimulus and control compartments, and within the bioassay compartment. *Melolontha* white grubs orientated to the CO_2 source, which corroborates earlier findings reported by Hasler (1986). Similar soil compartment systems have successfully been used to study the orientation of mycorrhizal fungi to their symbiotic roots (Vierheilig *et al.* 1998). Moving white grubs were often seen pushing their heads into the substrate or scraping substrate off the walls at different locations within their caves. Both behaviours may serve to probe the surroundings during host search.

However, neither roots of red clover (an accepted host) nor of dandelion (a preferred host) did attract *M. melolontha* larvae. The fact that Hasler (1986) found cut roots to attract *M. melolontha* might be due to the release of attractive chemicals from damaged roots. In the Exp. 2 and 3 with roots, the CO_2 concentration gradients were lower (Exp. 2 week 9, Exp. 3 week 9), in the same range (Exp. 3 week 11 and 13), or higher (Exp. 2 week 11 and 13) than in Exp. 1. But in any case, the gradients were at least one order of magnitude above the minimum required to attract white grubs (see results, and Hasler 1986). Moreover, CO_2-enriched air failed to attract white grubs when presented on the background of dandelion root exudates (Exp. 4). Thus, in Exp. 2–4, CO_2 was offered to the white grubs in a concentration which could be expected to elicit a positive response. However, no attraction was observed. Since Exp. 1 showed that the bioassay set-up is suitable to examine such a response, the non-attractiveness of CO_2 in the presence of plant roots or root exudates may be interpreted in such a way that root components interfere with or 'mask' the attractiveness of CO_2.

The concept of 'chemical masking' describes (a) the balancing of a resource's attractiveness by repellent compounds (e. g. Isaacs *et al.* 1993; Hori & Komatsu 1997; Held *et al.* 2003; Mauchline *et al.* 2005), or (b) the inhibition of an oriented response to an attractive stimulus by masking compounds that have no repellent activity *per se* (Yamasaki *et al.* 1997). No repellent effects were observed when offering roots or root exudates. Thus, we suggest that the attractiveness of CO_2 is not compensated by repellent components, but 'masked' by other, non-repellent cues released from the roots. Besides processing in the central nervous system, a physiological basis for the latter phenomenon is discussed in vertebrate olfactory systems as inhibition of responses to a stimulus by another compound at the peripheral level (Duchamp-Viret *et al.* 2003).

CO_2 did not function as foraging kairomone for cockchafer larvae on the background of root exudates or volatiles in our experiments. Therefore, the question arises in which context CO_2 sources might become relevant and attractive to white grubs (Hasler 1986, Exp. 1). In the upper soil layers, CO_2 emissions are ubiquitous. Sources may be plant roots, bacteria, fungi, conspecifics, or even predators. Furthermore, it diffuses rapidly through the pore system. Established gradients are horizontally (between roots) less pronounced than vertically. For these reasons, the suitability of CO_2 for host location in root-feeding insects has recently been questioned (Johnson & Gregory 2006). After hibernation, cockchafer larvae ascent from lower frost safe soil layers to feed close to the surface during the warm season (Niklas 1974). At this occasion, the orientation to the upper soil layers and to bulk plant roots may be guided by vertical CO_2 gradients.

Also above-ground herbivores are faced with the ubiquitous availability and presence of various sources of CO_2. Nevertheless, CO_2 has been shown a reliable semiochemical for them in different contexts. The moth *Cactoblastis cactorum* detects CO_2 gradients close to its host plant's surface to assess the vigour of the leaf before ovipositing (Stange *et al.* 1995). The Queensland fruit fly, *Batrocera tryoni*, detects small lesions in fruit skin by elevated CO_2 levels. The lesions facilitate insertion of the ovipositor into the fruit (Stange 1999). Foraging hawkmoths, *Manduca sexta*, choose flowers from *Datura wrightii*, which emit high amounts of CO_2. The flowers' CO_2 emissions are correlated to the amount of nectar available in them (Guerenstein *et al.* 2004; Thom *et al.* 2004). In haemotophagous insects, CO_2 may enhance the attractiveness of volatiles emanating from hosts (Barrozo & Lazzari 2005) or sensitize foraging individuals to host odours (Dekker *et al.* 2005). Thus, we suppose that CO_2 in context with other volatiles or specific cues may convey reliable information to the receiver and modulate the receiver´s behaviour. Whether this is true also for below-ground herbivores needs to be elucidated by future investigations.

The results presented here show that *M. melolontha* larvae need more chemical stimuli than CO_2 alone to localize the host. Whether plant roots are active players in 'hiding' their CO_2 emissions by masking compounds has to be answered in future projects focusing particularly on the chemical analysis of root exudates modifying *M. melolontha* larval behaviour.

8.5 Acknowledgements

Hans-Peter Haschke, Freie Universität Berlin, Inst. of Biology, Plant Physiology gave helpful advice regarding the rearing of the plants and the choice of the substrate. Nana Hesler, Christian Joppich, and Mathieu T'Flachebba helped with the bioassays. We thank Ute Braun and Renate Jonas for rearing the larvae. This work was financed by the Deutsche Forschungsgemeinschaft (DFG Hi 416/13).

8.6 References

Arnon DI & Hoagland DR (1940). Crop production in artificial culture solutions and in soils with special reference to factors influencing yields and absorption of inorganic nutrients. Soil Sci 50:463-483.

Barrozo RB & Lazzari CR (2005). Orientation behaviour of the blood-sucking bug *Triatoma infestans* to short-chain fatty acids: Synergistic effect of L-lactic acid and carbon dioxide. Chem Senses 29:833-841.

Ben Yakir D, Goldberg AM & Chen M (1995). Laboratory efficacy screening of insecticides for control of *Maladera matrida* larvae. Phytoparasitica 23:119-25.

Bengtsson G, Hedlung K & Rundgren S (1991). Selective odor perception in the soil collembola *Onychiurus armatus*. J Chem Ecol 17:2113-2125.

Blanco Montero CA & Hernandez G (1995). Mechanical control of white grubs (Coleoptera: Scarabidae) in turfgrass using aerators. Environ Entomol 24:243-5.

Crocker RL, Marshall D & Kubica-Breier JS (1990). Oat, wheat, and barley resistance to white grubs of *Phyllophaga congrua* (Coleoptera: Scarabaeidae). J Econ Entomol 83:1558-1562.

Crutchfield BA, Potter DA & Powell AJ (1995). Irrigation and nitrogen fertilization effects on white grub injury to Kentucky bluegrass and tall fescue turf. Crop Sci 35:1122-1126.

Dekker T, Geier M & Cardé RT (2005). Carbon dioxide instantly sensitizes female yellow fever mosquitoes to human skin odours. J Exp Biol 208:2963-2972.

Duchamp-Viret P, Duchamp A & Chaput MA (2003). Single olfactory sensory neurons simultaneously integrate the components of an odour mixture. Europ J Neurosci 18:2690-2696.

Ene IM (1942). Experimentaluntersuchungen über das Verhalten des Maikäferengerlings (*Melolontha* spec.). Z Ang Entomol 26:529-600.

Facundo HT, Linn Jr. CE, Villani MG & Roelofs WL (1999). Emergence, mating, and postmating behaviors of the Oriental beetle (Coleoptera: Scarabaeidae). J Ins Behav 12:175-192.

Guerenstein PG, Yepez EA, van Haaren Y, Williams DG & Hilderbrand JG (2004). Floral CO_2 emission may indicate food abundance to nectar-feeding moths. Naturwissenschaften 91:329-333.

Harari AR, Ben-Yakir D & Rosen D (2000). Male pioneering as a mating strategy: The case of the beetle *Maladera matrida*. Ecol Entomol 25:387-394.

Hasler T (1986). Abundanz- und Dispersionsdynamik von *Melolontha melolontha* (L.) in Intensivobstanlagen. Dissertation. Eidgenössische Technische Hochschule, Zürich.

Hauss R (1975). Methoden und erste Ergebnisse zur Bestimmung der Wirtspflanzen des Maikäferengerlings (*Melolontha melolontha* L.). Mitt Biol Bundesanst Land- & Forstw 163:72-77.

Hauss R & Schütte F (1976). Zur Polyphagie der Engerlinge von *Melolontha melolontha* L. an Pflanzen aus Wiese und Ödland. Anz Schädlingsk Pflanzensch Umweltsch 49:129-132.

Held DW, Gosinska P & Potter DA (2003). Evaluating companion planting and non-host masking odors for protecting roses from the Japanese beetle (Coleoptera : Scarabaeidae). J Econ Entomol 96:81-87.

Hori M & Komatsu H (1997). Repellency of rosemary oil and its components against the onion aphid, *Neotoxoptera formosana* (TAKAHASHI) (Homoptera, Aphididae). Appl Entomol Zool 32:303-310.

Hurpin B (1962). Famille des Scarabaeides. In: Balachowsky AS (ed). Entomologie appliquée à l'agriculture. Masson et Cie, Paris, pp 24–203.

Isaacs R, Hardie J, Hick AJ, Pye BJ, Smart LE, Wadhams LJ & Woodcock CM (1993). Behavioral responses of *Aphis fabae* to isothiocyanates in the laboratory and field. Pest Sci 39:349-355.

Johnson SN, Gregory PJ, Greenham JR, Zhang X & Murray PJ (2005). Attractive properties of an isoflavonoid found in white clover root nodules on the colver root weevil. J Chem Ecol 31:2223-2229.

Johnson SN & Gregory PJ (2006). Chemically-mediated host-plant location and selection by root-feeding insects. Physiol Entomol 31:1-13.

Keller S (1986). Biologie und Populationsdynamik. In: Neuere Erkenntnisse über den Maikäfer. Beih Thurg Naturforsch Ges, Frauenfeld. 12-39.

Keller S, Schweizer S, Keller E & Brenner H (1997). Control of white grubs (*Melolontha melolontha* L.) by treating adults with the fungus *Beauveria brongniartii*. Biocon Sci Techn 7:105-116.

Kim JY & Leal WS (1999). Eversible pheromone glands in a Melolonthine beetle, *Holotrichia parallela*. J Chem Ecol 25:825-833.

Klingler J (1958). Die Bedeutung der Kohlendioxid-Ausscheidung der Wurzeln für die Orientierung der Larven von *Otiorrhynchus sulcatus* F. und anderer bodenbewohnender phytophager Insektenarten. Mitt Schweiz Entomol Ges 3-4:206-269.

Koppenhoefer AM & Kaya HK (1997). Additive and synergistic interaction between entomopathogenic nematodes and *Bacillus thuringiensis* for scarab grub control. Biol Con 8:131-137.

Koppenhoefer AM, Choo HY, Kaya HK, Lee DW & Gelernter WD (1999). Increased field and greenhouse efficacy against scarab grubs with a combination of an entomopathogenic nematode and *Bacillus thuringiensis*. Biol Con 14:37-44.

Larsson MC, Stensmyr MC, Bice SB & Hansson BS (2003). Attractiveness of fruit and flower odorants detected by olfactory receptor neurons in the fruit chafer *Pachnoda marginata*. J Chem Ecol 29:1253-1268.

Leal WS (1998). Chemical ecology of phytophagous scarab beetles. Annu Rev Entomol 43:39-61.

Mateille T (1994). Biologie de la relation plantes-nématodes: Perturbations physiologiques et mécanismes de défense des plantes. Nematologica 40:276-311.

Mauchline AL, Osborne JL, Martin AP, Poppy GM & Powell W (2005). The effects of non-host plant essential oil volatiles on the behaviour of the pollen beetle *Meligethes aeneus*. Ent Exp App 114:181-183.

Niklas OF (1974). Familienreihe Lamellicornia, Blatthornkäfer. In: Schwenke W (ed). Die Forstschädlinge Europas. Parey, Hamburg und Berlin, pp 85-129.

Park SY, Kim JS, Lee JB, Esler MB, Davis RS & Wielgosz RI (2004). A redetermination of the argon content of air for buoyancy corrections in mass standard comparisons. Metrologia 41:387-395.

Rasmann S, Köllner TG, Degenhard J, Hiltpold I, Toepfer S, Kuhlmann U, Gershenzon J & Turlings TCJ (2005). Recruitment of entomopathogenic nematodes by insect damaged maize roots. Nature 434:732-737.

Reinecke A, Ruther J & Hilker M (2002a). The scent of food and defence: Green leaf volatiles and toluquinone as sex attractant mediate mate finding in the European cockchafer *Melolontha melolontha*. Ecol Lett 5:257-263.

Reinecke A, Ruther J & Hilker M (2005). Electrophysiological and behavioural response of *Melolontha melolontha* to saturated and unsaturated aliphatic alcohols. Entomol Exp Appl 115:33-40.

Rogers ME & Potter DA (2002). Kairomones from scarabaeid grubs and their frass as cues in below-ground host location by the parasitoids *Tiphia vernalis* and *Tiphia pygidialis*. Entomol Exp Appl 102:307-314.

Ruther J, Reinecke A, Tolasch T & Hilker M (2001). Make love not war: A common arthropod defence compound as sex pheromone in the forest cockchafer *Melolontha hippocastani*. Oecologia 128:44-47.

Ruther J, Reinecke A & Hilker M (2002). Plant volatiles in the sexual communication of *Melolontha hippocastani*: Response towards time-dependent bouquets and novel function of (Z)-3-hexen-1-ol as a sexual kairomone. Ecol Entomol 27:76-83.

Ruther J & Mayer CJ (2005). Response of garden chafer, *Phyllopertha horticola*, to plant volatiles: From screening to application. Ent Exp App 115:51-59.

Sachs L (1992). Angewandte Statistik, 7th edn. Springer Verlag, Berlin.

Schwerdtfeger F (1939). Untersuchungen über die Wanderungen des Maikäfer-Engerlings (*Melolontha melolontha* L. und *Melolontha hippocastani* F.). Z Ang Entomol 26:215-252.

Sheppard SK & Lloyd D (2002). Diurnal oscillations in gas production (O_2, CO_2, CH_4, and N_2) in soil monoliths. Biol Rhythm Res 33:577-591.

Stange G, Monro J, Stowe S & Osmond CB (1995). The CO_2 sense of the moth *Cactoblastis cactorum* and its probable role in the biological control of the CAM plant *Opuntia stricta*. Oecologia 102:341-352.

Stange G (1999). Carbon dioxide is a close-range oviposition attractant in the Queensland fruit fly *Bactrocera tryoni*. Naturwissenschaften 86:190-192.

Sutherland ORW & Hillier JR (1974). Olfactory response of *Costelytra zealandica* (Coleoptera: Melolonthinae) to the roots of several pasture plants. New Zeal J Zool 1:365-369.

Sutherland ORW, Mann J & Hillier JR (1975a). Feeding deterrent for the grass grub *Costelytra zealandica* (Coleoptera: Scarabaeidae) in the root of a resistant pasture plant, *Lotus pedunculatus*. New Zeal J Zool 2:509-512.

Sutherland ORW, Hood ND & Hillier JR (1975b). Lucerne root saponins a feeding deterrent for the grass grub, *Costelytra Zealandica* (Coleoptera: Scarabaeidae). New Zeal J Zool 2:93-100.

Thom C, Guerenstein PG, Mechaber WL & Hildebrand JG (2004). Floral CO_2 reveals flower profitability to moths. J Chem Ecol 30:1285-1288.

Van Tol RWHM, van der Sommen ATC, Boff MIC, van Bezooijen J, Sabelis MW & Smits PH (2001). Plants protect their roots by alerting the enemies of grubs. Ecol Lett 4:292-294.

Vierheilig H, Alt-Hug M, Engel-Streitwolf R, Mäder P & Wiemken A (1998). Studies on the attractional effect of root exudates on hyphal growth of an arbuscular mycorrhizal fungus in a soil compartment-membrane system. Plant and Soil 203:137-44.

Wei X, Xu X & Deloach CJ (1995). Biological control of white grubs (Coleoptera: Scarabaeidae) by larvae of *Promachus yesonicus* (Diptera: Asilidae) in China. Biol Con 5:290-296.

Wensler RJ & Dudzinski AE (1971). Gustatory discrimination between plants by larvae of the scarabaeid *Sericesthis geminata* (Coleoptera). Entomol Exp Appl 14:441-448.

Williams RN, Dickle DS, McGovern TP & Klein MG (2000). Development of an attractant for the scarab pest *Macrodactylus subspinosus* (Coleoptera: Scarabaeidae). J Econ Entomol 93:1880-1884.

Yamasaki T, Sato M & Sakoguchi H (1997). (–)–Germacrene D: Masking substance of attractants for the cerambycid beetle, *Monochamus alternatus* (HOPE). Appl Entomol Zool 32:423-429.

Zheng XY & Sinclair JB (1996). Chemotactic response of *Bacillus megaterium* strain B153-2-2 to soybean root and seed exudates. Physiol Mol Plant Pathol 48:21-35.

Chapter 9

General discussion

Overall, the studies presented in this thesis show that life of *M. melolontha* is essentially guided by chemicals (pheromones and kairomones) both above and below ground. The various specific aspects are discussed in chapters 2–8. Here, some considerations of so far unpublished data are combined with the published data. Furthermore, based on these summarizing aspects and conclusions some urgent future studies are suggested.

9.1 Mate finding

M. melolontha males have been shown to locate potential mates using female-feeding induced green leaf alcohols as primary sex attractants (Chap. 2, Chap. 5, and Chap. 6). Thus, volatiles from the first trophic level, the host plant, are beneficial to the second trophic level as they promote mate finding. Based on the definition of kairomones given by Dicke and Sabelis (1988), and in analogy to the term 'aggregation kairomone' introduced by Loughrin *et al.* (1995), Ruther *et al.* (2002a) denominated semiochemicals mediating mate finding across trophic levels as 'sexual kairomone'. The data presented here support the validity of this concept.

The question arises whether male cockchafers are at least partially independent in their mate finding strategy from female pheromone signalling. Experiments addressing this question should test at which conditions males behaviourally respond to leaf alcohols and female-derived attractants. Their responsiveness might depend on their physiological condition or on their experience as well as on abiotic factors. Furthermore, the males´ behavioural response might depend on the female pheromone release patterns (see also below).

The attractiveness of leaf alcohols to males is enhanced by the beetle-derived attractants toluquinone and phenol (Chap. 2, Chap. 3, and Chap. 7). It is the first time that an insect sex pheromone, namely toluquinone, has been identified, which is not attractive *per se* (Reddy & Guerrero 2004). However, the function of phenol needs a closer discussion. This compound has been identified in female extracts. It is attractive to *Melolontha* males by itself and enhances the attractiveness of the green leaf alcohol (*Z*)-3-hexen-1-ol (Chap. 3). However, in ternary mixtures, funnel trap catch numbers drop when high amounts of phenol are added to the most attractive ratio of (*Z*)-3-hexen-1-ol and benzoquinone (5 mg : 0.5 mg). In contrast, in binary mixtures of (*Z*)-3-hexen-1-ol and phenol, the highest tested doses of phenol were most attractive in both, forest and European cockchafers (Chap. 7). These results are difficult to understand. The different female-derived volatiles (toluquinone and phenol) that attract males might be released in dependence of the female´s age, her mating status, or other factors such as symbiotic microorganisms that might be involved in the biosynthesis of the attractive components.

It is known from other scarab beetles that the release of sex pheromones varies with time, diurnal rhythm, the mating status, or other unknown factors (Ladd 1970a; Leal 1993; Yarden *et al.* 1994; Heath *et al.* 2002; Kim *et al.* 2002). Furthermore, some scarab sex pheromones are known to be produced, and thus controlled by symbiontic bacteria (Hoyt & Osborne 1971). In other insects, pheromone production or release may require the presence of specific hosts and signalling may be delayed until this host is met (Landolt & Phillips 1997; Reddy & Guerrero 2004). In *Popillia japonica*, a scarab with life history traits comparable to *M. melolontha* regarding the delay between emergence and egg deposition, a last male sperm advantage is documented, even though the female sex pheromone release drops after the first mating (Ladd 1970b; Van Timmerman *et al.* 2000). *M. melolontha* females start mating soon after emergence, and may mate several times before ovipositing (Hurpin 1962; Niklas 1974). On the other hand, they migrate during the season from forest edges with male biased sex ratios into the forest where the sex ratio is female biased (Vogel 1956), which might be interpreted as an escape from a male dominated area.

Accordingly future experiments should address the following questions: (I) Does the composition of *M. melolontha* female headspace volatiles vary with (a) time, (b) mating status, (c) feeding status. (II) Where are the attractants produced? Are they produced in the same tissues? Are the attractants produced by enzymes of the females or of symbiotic microorganisms? (III) Does the first or the last male win the sperm competition, or do females exert control on the fathering of their offspring?

Behavioural experiments revealed that *Melolontha* males discriminate conspecific from heterospecific females (Chap. 4). Intraspecific acoustical communication and optical cues may be excluded as discussed in Chap. 4. Since full body extracts do not necessarily reflect release rates, analyses of *M. melolontha* and *M. hippocastani* headspace volatile compositions would shed light on the underlying chemical signals. However, the experimental data show that different pheromone blends alone would not account for precopulatory species isolation in sympatric populations of *Melolontha* species. When given the choice between heterospecific males and females, males significantly preferred the females instead of continuing their search for conspecific females, which were freely available in the surrounding of the test. Copulation in *Melolontha* beetles lasts for hours (Krell 1992) and may therefore be considered a serious male investment in time. Thus, other mechanisms should contribute to reduce the probability of "wrong" matings, but remain to be elucidated.

9.2 Orientation of females

Female antennae respond in electroantennographic experiments to almost all compounds perceived by male antennae (Chap. 2, Chap. 5). Male antennal responses to certain leaf alcohols are stronger than responses from female antennae (Chap. 6). This sexual dimorphism in the perception of leaf alcohols highlights the importance of these compounds to cockchafer males. However, dose-dependent responses to any tested compound were recorded from female antennae as well (Chap. 6). Thus, female perception of green leaf volatiles may not be as sensitive as male perception, but allows in either case to behaviourally discriminate between different components and concentrations.

During field experiments, females have been observed flying before the main swarming flight at dusk started. At these times funnel traps were already placed, but preferences of the females for any of the tested stimuli were never detected. Beyond the experiments reported here, trials were undertaken to elucidate female olfactory orientation in specific contexts.

- In the melolonthine species, *Maladera matrida*, trap colour modulated the attractiveness of odour lures to females, but not to males (Falach & Shani 2000). After emergence, European cockchafers fly to nearby forest edges, where they search for host trees and mates (Niklas 1974; Keller 1986a). In a funnel trap bioassay using standard (grey) and yellow traps during the post-emergence flights to the forest edges, females did not respond to a GLV mixture, which was highly attractive to males of the species, irrespective of the trap colour.

- Male *M. matrida* locate host plants and start feeding upon them. Females locate these plants using male-feeding induced volatiles (Harari *et al.* 2000). Feeding-induced leaf volatiles also serve as aggregation kairomones in the Japanese beetle, *Popilllia japonica* (Loughrin *et al.* 1995, 1996a). Cockchafer males emerge about two days earlier than females (Niklas 1974). Antennae from both sexes respond to methyl salicylate (Chap. 5), a compound typically released by leaves of many plant species after insect feeding. However, females did not respond to this compound when it was applied in funnel traps.

- Young first instar cockchafer larvae feed on poaceaen roots. Second and third instars prefer dandelion roots as host (Hauss 1975; Hauss & Schütte 1976; Keller 1986b). During egg deposition flights, intact and damaged plant material from dandelion (*Taraxacum* sect. *ruderalia*) and mixtures of Poaceae was used as odour source in traps at 0 cm, 50 cm, and 150 cm height. Again, females did not respond to the stimuli.

So far, these observations and the presented data do not allow conclusions about female olfactory orientation. Males locate females, which represent, as most traps do, spot-like odour sources. Females must find host trees to feed on. Furthermore, they need to locate oviposition sites suitable for larval development. These feeding and oviposition sites might be found by more diffuse, and other than spot-like odour sources. Future investigations regarding female cockchafer orientation must take these considerations into account.

9.3 Orientation of larvae in the soil

A new soil arena type has been developed to investigate responses from *M. melolontha* larvae to chemical stimuli from plant roots (Chap. 8). Divided into three compartments, the arena allowed (I) an almost undisturbed development of the plant roots even during the experiments, (II) the yield of root exudates without disturbance of the rhizosphere and without destruction of the roots, (III) to monitor CO_2 concentrations without disturbance of the experiment, and (VI) the observation of white grubs without physical access to the plant roots, but exposed to chemical stimuli from the plant and a control compartment. CO_2 is a known attractant to white grubs

(Klingler 1956; Hasler 1986). Thus, the suitability of the new soil arena has been tested and demonstrated with CO_2. White grubs oriented to the test compartment when synthetic CO_2 was used as stimulus, but not when plant roots were offered, although the roots generated adequate CO_2 concentrations. The response to CO_2 was also suppressed on the background of dandelion root exudates. Arrestment of the larvae in the vicinity of host plant roots or masking of the attractiveness of CO_2 by root-derived chemicals may be causes for the modified behaviour. Behavioural observations are in favour of the 'masking-hypothesis', although future experiments are needed to draw final conclusions (Chap. 8). It may seem surprising that white grubs did not locate preferred host plants in a choice test. However, comparable findings are known from other scarab species above and below ground (Sutherland *et al.* 1974, 1975a,b; Loughrin *et al.* 1995, 1996a,b, 1997). From the plant´s perspective, hiding between other plant roots would potentially reduce the costs of herbivory. The newly developed arena offers a variety of options to further investigate the orientation of *Melolontha* white grubs and, adapted in size, also of other root-feeding larvae. Gauze or PTFE membranes may be used to delimit the compartments of the arena, thereby selectively offering gases and dissolved compounds or gases only as stimulus. Always excluding tested animals from physical contact to the tested roots, stimuli may be generated from controlled herbivory or mechanical damage of roots, and exudates may be extracted for bioassays and chemical analysis without any root manipulation.

9.4 References

Butenandt A, Beckmann R, Stamm D & Hecker E (1959). Über den Sexuallockstoff des Seidenspinners *Bombyx mori*. Reindarstellung und Konstittution. Z Naturforschung 14b:283-284.

Dicke M & Sabelis MW (1988). Infochemical terminology: Based on cost-benefit analysis rather than origin of compounds? Funct Ecol 2:131-139.

Ene IM (1942). Experimentaluntersuchungen über das Verhalten des Maikäferengerlings (*Melolontha* spec.). Z Ang Entomol 26:529-600.

Falach L & Shani A (2000). Trapping efficiency and sex ratio of *Maladera matrida* beetles in yellow and black traps. J Chem Ecol 26:2619-2624.

Hansson BS, Larsson MC & Leal WS (1999). Green leaf volatile-detecting olfactory receptor neurones display very high sensitivity and specifity in a scarb beetle. Physiol Entomol 24:121-126.

Harari AR, Ben-Yakir D & Rosen D (2000). Male pioneering as a mating strategy: The case of the beetle *Maladera matrida*. Ecol Entomol 25:387-394.

Hasler T (1986). Abundanz- und Dispersionsdynamik von *Melolontha melolontha* (L.) in Intensivobstanlagen. Dissertation. Eidgenössische Technische Hochschule, Zürich

Hauss R (1975). Methoden und erste Ergebnisse zur Bestimmung der Wirtspflanzen des Maikäferengerlings (*Melolontha melolontha* L.). Mitt Biol Bundesanst Land & Forstw 163:72-77.

Hauss R & Schütte F (1976). Zur Polyphagie der Engerlinge von *Melolontha melolontha* L. an Pflanzen aus Wiese und Ödland. Anz Schädlingsk Pflanzensch Umweltsch 49:129-132.

Heath JJ, William RN & Phelan PL (2002). Aggregation and male attraction to feeding virgin females in *Macrodactylus subspinosus* (F.) (Coleoptera: Scarabaeidae: Melolonthinae). Environ Entomol 31:934-940.

Henzell RF & Lowe MD (1970). Sex attractant of the gras grub beetle. Science 168:1005-1006.

Hoyt CP & Osborne GO (1971). Production of an insect sex attractant by symbiotic bacteria. Nature 230:472-473.

Hurpin B (1962). Famille des Scarabaeides. In: Balachowsky AS (ed). Entomologie appliquée à l'agriculture. Masson et Cie, Paris, pp 24–203.

Johnson SN, Gregory PJ, Greenham JR, Zhang X & Murray PJ (2005). Attractive properties of an isoflavonoid found in white clover root nodules on the clover root weevil. J Chem Ecol 31:2223-2229.

Johnson SN & Gregory PJ (2006). Chemically-mediated host-plant location and selection by root-feeding insects. Physiol Entomol 31:1-13.

Karg G & Suckling M (1999). Applied aspects of insect olfaction. In: Hansson BS (ed). Insect Olfaction. Springer, Berlin Heidelberg, pp 351-377.

Keller E (1986). Chemische Maikäferbekämpfung. In: Neuere Erkenntnisse über den Maikäfer. Beih Mitt Thurg Naturforsch Ges, Frauenfeld, pp 69-75.

Keller S (1986a). Biologie, Populationsdynamik. In: Neuere Erkenntnisse über den Maikäfer. Beih Mitt Thurg Naturforsch Ges, Frauenfeld, pp 13-40.

Keller S (1986b). Historischer Rückblick, Kulturmassnahmen. In: Neuere Erkenntnisse über den Maikäfer. Beih Mitt Thurg Naturforsch Ges, Frauenfeld, pp 61-68.

Kim JY, Hasegawa M & Leal WS (2002). Individual variation in pheromone emission and termination patterns in female *Anomala cuprea*. Chemoecology 12:121-124.

Klingler J (1958). Die Bedeutung der Kohlendioxid-Ausscheidung der Wurzeln für die Orientierung der Larven von *Otiorrhynchus sulcatus* F. und anderer bodenbewohnender phytophager Insektenarten. Mitt Schweiz Entomol Ges 3-4:206-269.

Krell FT & Fery H (1992). Familienreihe Lamellicornia. In: Lohse GA & Lucht WH (eds). Die Käfer Mitteleuropas. Goecke & Evers, Krefeld, pp 200-252.

Krell FT (2004). Bestimmung von Larven und Imagines der mitteleuropäischen *Melolontha*-Arten (Coleoptera: Scarabaeoidea). Laimburg J 1:211-219.

Ladd Jr. TL (1970a). Sex attraction in the Japanese beetle. J Econ Entomol 63:905-908.

Ladd Jr. TL (1970b). Mating competitiveness of male Japanese beetles sterilized with tepa. J Econ Entomol 63:438-439.

Landolt PJ & Philips TW (1997). Host plant influences on sex pheromone behavior of phytophagous insects. Annu Rev Entomol 42:371-391.

Larsson MC, Leal WS & Hansson BS (2001). Olfactory receptor neurons detecting plant odours and male volatiles in *Anomala cuprea* beetles (Coleoptera: Scarabaeidae). J Ins Physiol 47:1065-1076.

Leal WS, Sawada M, Matsuyama S, Kuwahara Y & Hasegawa M (1993). Unusual periodicity of sex pheromone production in the large black chafer *Holotrichia parallela*. J Chem Ecol 19:1381-1391.

Leal WS (1998). Chemical ecology of phytophagous scarab beetles. Annu Rev Entomol 43:39-61.

Leal WS, Oehlschlager AC, Zarbin PHG, Hidalgo E, Shannon PJ, Murata Y, Gonzales LM, Andrade R & Ono M (2003). Sex pheromone of the scarab beetle *Phyllophaga elenans* and some intriguing minor components. J Chem Ecol 29:15-25.

Loughrin JH, Potter DA & Hamilton-Kemp TR (1995). Volatile compounds induced by herbivory act as aggregation kairomones for the Japanese beetle (*Popillia japonica* NEWMAN). J Chem Ecol 21:1457-1467.

Loughrin JH, Potter DA, Hamilton-Kemp TR & Byers ME (1996a). Role of feeding-induced plant volatiles in aggregative behavior of the Japanese beetle (Coleoptera:Scarabaeidae). Environ Entomol 25:1188-1191.

Loughrin JH, Potter DA, Hamilton-Kemp TR & Byers ME (1996b). Volatile compounds from crabapple (*Malus* spp.) cultivars differing in susceptibility to the Japanese beetle (*Popillia japonica* NEWMAN). J Chem Ecol 22:1295-1305.

Loughrin JH, Potter DA, Hamilton-Kemp TR & Byers ME (1997). Response of Japanese Beetles (Coleoptera:Scarabaeidae) to leaf volatiles of susceptible and resistant maple species. Environ Entomol 26:334-342.

Mateille T (1994). Biologie de la relation plantes-nématodes: Perturbations physiologiques et mécanismes de défense des plantes. Nematologica 40:276-311.

Niklas OF (1974). Familienreihe Lamellicornia, Blatthornkäfer. In: Schwenke W (ed). Die Forstschädlinge Europas. Parey, Hamburg und Berlin, pp 85-129.

Nojima S, Robbins PS, Salsbury GA, Morris BD, Roelofs WL & Villani MG (2003a). 1-Leucine methyl ester: The female-produced sex pheromone of the scarab beetle, *Phyllophaga lanceolata*. J Chem Ecol 29:2439-2446.

Nojima S, Sakata T, Yoshimura K, Robbins PS, Morris BD & Roelofs WL (2003b). Male-specific EAD active compounds produced by female European chafer *Rhizotrogus majalis* (RAZOUMOWSKY). J Chem Ecol 29:503-507.

Rasmann S, Köllner TG, Degenhard J, Hiltpold I, Toepfer S, Kuhlmann U, Gershenzon J & Turlings TCJ (2005). Recruitment of entomopathogenic nematodes by insect damaged maize roots. Nature 434:732-737.

Reddy GVP & Guerrero A (2004). Interaction of insect pheromones and plant semiochemicals. Trends Plant Sci 9:253-261.

Reinecke A, Ruther J & Hilker M (2002). The scent of food and defence: Green leaf volatiles and toluquinone as sex attractant mediate mate finding in the European cockchafer *Melolontha melolontha*. Ecol Lett 5:257-263.

Renou M & Guerrero A (2000). Insect parapheromones in olfaction research and semiochemical-based pest control strategies. Annu Rev Entomol 45:605-630.

Rochat D, Morin JP, Kakul T, Beaudouin-Ollivier L, Prior R, Renou M, Malosse I, Stathers T, Embupa S & Laup S (2002). Activity of male pheromone of Melanesian rhinoceros beetle *Scapanes australis*. J Chem Ecol 28:479-500.

Ruther J, Reinecke A, Thiemann K, Tolasch T, Francke W & Hilker M (2000). Mate finding in the forest cockchafer, *Melolontha hippocastani*, mediated by volatiles from plants and females. Physiol Entomol 25:172-9.

Ruther J, Reinecke A, Tolasch T & Hilker M (2001). Make love not war: A common arthropod defence compound as sex pheromone in the forest cockchafer *Melolontha hippocastani*. Oecologia 128:44-47.

Ruther J, Reinecke A & Hilker M (2002a). Plant volatiles in the sexual communication of *Melolontha hippocastani*: Response towards time-dependent bouquets and novel function of (*Z*)-3-hexen-1-ol as a sexual kairomone. Ecol Entomol 27:76-83.

Ruther J, Meiners T & Steidle JLM (2002b). Rich in phenomena – lacking in terms: A classification of kairomones. Chemoecology 12:161-167.

Ruther J & Hilker M (2003). Attraction of forest cockchafer *Melolontha hippocastani* to (*Z*)-3-hexen-1-ol and 1,4-benzoquinone: Application aspects. Entomol Exp Appl 107:141-147.

Schlyter F & Birgersson GA (1999). Forest Beetles. In: Hardie J & Minks AK (eds). Pheromones of Non-Lepidopteran Insects Associated with Agricultural Plants. CABI publishing, Wallingford, pp 113-148.

Schwerdtfeger F (1939). Untersuchungen über die Wanderungen des Maikäfer-Engerlings (*Melolontha melolontha* L. und *Melolontha hippocastani* F.). Z Ang Entomol 26:215-252.

Stensmyr MC, Larsson MC, Bice S & Hansson BS (2001). Detection of fruit- and flower-emitted volatiles by olfactory receptor neurons in the polyphagous fruit chafer *Pachnoda marginata* (Coleoptera: Cetoniinae). J Comp Physiol A 187:509-519.

Sutherland ORW & Hillier JR (1974). Olfactory response of *Costelytra zealandica* (Coleoptera: Melolonthinae) to the roots of several pasture plants. New Zeal J Zool 1:365-369.

Sutherland ORW, Mann J & Hillier JR (1975a). Feeding deterrent for the grass grub *Costelytra zealandica* (Coleoptera: Scarabaeidae) in the root of a resistant pasture plant, *Lotus pedunculatus*. New Zeal J Zool 2:509-512.

Sutherland ORW, Hood ND & Hillier JR (1975b). Lucerne root saponins a feeding deterrent for the grass grub, *Costelytra Zealandica* (Coleoptera: Scarabaeidae). New Zeal J Zool 2:93-100.

Tamaki Y, Sugie H & Noguchi H (1985). Methyl (*Z*)-5-tetradecenoate: Sex-attractant pheromone of the soybean beetle, *Anomala rufocuprea* MOTSCHULSKY (Coleoptera: Scarabaeidae). Appl Entomol Zool 20:359-361.

Tolasch T, Sölter S, Tóth M, Ruther J & Francke W (2003). (R)-Acetoin – female sex pheromone of the summer chafer *Amphimallon solstitiale* (L.). J Chem Ecol 29:1045-1050.

Tóth M, Subchev M, Sredkov I, Szarukan I & Leal WS (2003). A sex attractant for the scarab beetle *Anomala solida* ER. J Chem Ecol 29:1643-1649.

Tumlinson JH, Klein MG, Dolittle RE, Ladd Jr. TL & Proveaux AT (1977). Identification of the female Japanese beetle sex pheromone: Inhibition of the male response by an enantiomer. Science 197:789-792.

Van Timmermann SJ, Switzer PV & Kruse KC (2000). Emergence and reproductive patterns in the Japanese beetle, *Popillia japonica* (Coleoptera: Scarabaeidae). J Kansas Entomol Soc 74:17-27.

Van Tol RWHM, van der Sommen ATC, Boff MIC, van Bezooijen J, Sabelis MW & Smits PH (2001). Plants protect their roots by alerting the enemies of grubs. Ecol Lett 4:292-294.

Vierheilig H, Alt-Hug M, Engel-Streitwolf R, Mäder P & Wiemken A (1998). Studies on the attractional effect of root exudates on hyphal growth of an arbuscular mycorrhizal fungus in a soil compartment-membrane system. Plant Soil 203:137-44.

Vogel W (1956). Entmischungen innerhalb der Maikäferpopulationen im Zusammenhang mit dem Wandern der Käfer ins Waldesinnere. Z Ang Entomol 38:206-216.

Ward A, Moore C, Anitha V, Wightman J & Rogers DJ (2002). Identification of the sex pheromone of *Holotrichia reynaudi*. J Chem Ecol 28:515-522.

Yarden G & Shani A (1994). Evidence for volatile chemical attractants in the beetle *Maladera matrida* ARGAMAN (Coleoptera:Scarabaeidae). J Chem Ecol 20:2673-2685.

Zheng XY & Sinclair JB (1996). Chemotactic response of *Bacillus megaterium* strain B153-2-2 to soybean root and seed exudates. Physiol Mol Plant Pathol 48:21-35.

Zweigelt F (1928). Der Maikäfer. Monogr Ang Entomol 9:1-446.

Chapter 10

Summary - Zusammenfassung

10.1 English summary

European cockchafers, *Melolontha melolontha* L. (Coleoptera: Scarabaeidae: Melolonthinae), are a severe pest in agriculture and horticulture when calamitous mass breeding occurs. The polyphagous larvae feed upon plant roots, while the adults may heavily damage above-ground foliage. As a pest, the species has extensively been studied in many biological aspects. One option to control a herbivore pest insect is the use of those naturally occurring chemicals that mediate its sexual communication or its location of the host plant. However, in the beginning of the studies presented here hardly anything was known on the role of chemical cues involved in *M. melolontha* mate finding and host plant location. The aim of this chemo-ecological study was to elucidate this lack of knowledge by laboratory and field experiments coupled with chemical analyses of pheromones and kairomones.

10.1.1 Mate finding and pheromone-related communication

Cockchafers perform a spectacular swarming flight around host trees at dusk. Counting of flying and resting beetles revealed that males exclusively perform the swarming flight while females remain on the host tree leaves they feed or have fed upon. In a landing cage bioassay conducted in the field, swarming cockchafer males preferred cages baited with females to cages baited with males. Gas chromatographic analysis of beetle extracts with electroantennographic detection revealed the presence of electrophysiologically active compounds, among them toluquinone, phenol, and 1,4-benzoquinone, the sex pheromone of the closely related forest cockchafer, *M. hippocastani* FABR. In funnel trap bioassays none of the quinones is attractive to males *per se*. Volatiles from mechanically damaged leaves and a mixture of green leaf volatiles (GLV) mimicking the bouquet of mechanically damaged leaves are highly attractive to *M. melolontha* males. The attractiveness of the same GLV mixture is synergistically enhanced when toluquinone is added to the lure. In the same setup, 1,4-benzoquinone is behaviourally not active. Thus, based on a sexual dimorphism in flight behaviour, GLV act as sexual kairomones and attract males to sites of female feeding damage. Toluquinone as a sex pheromone indicates that conspecific females are actually present, and synergistically enhances the attractiveness of the GLV. This constitutes the first report about an insect sex pheromone not being attractive on its own, but needing the concomitant presence of host plant volatiles to attract males to potential mates.

Phenol has been identified in female full body extracts from both, *M. hippocastani* and *M. melolontha*. In the field, phenol attracts males of both species and enhances the attraction of cockchafer males to the green leaf alcohol (*Z*)-3-hexen-1-ol. If equal ratio mixtures are compared, a mixture of phenol plus (Z)-3-hexen-1-ol is less attractive for *M. hippocastani* males than a mixture of (*Z*)-3-hexen-1-ol plus 1,4-benzoquinone, whereas phenol plus (*Z*)-3-hexen-1-ol attracts as many *M. melolontha* males as a mixture of (*Z*)-3-hexen-1-ol plus toluquinone. In both species three component mixtures containing phenol, (*Z*)-3-hexen-1-ol, and the respective benzoquinone in equal proportions do not capture more males than two component mixtures consisting of only (*Z*)-3-hexen-1-ol and the benzoquinone. These results show that phenol is another male attractant common to *M. hippocastani* and *M. melolontha*. However, when optimized ratios of (*Z*)-3-hexen-1-ol and the respective benzoquinone are used, addition of phenol reduces numbers of attracted males in both species. The exact function of phenol remains to be elucidated, and the term male attractant is used instead of sex pheromone.

Since toluquinone, 1,4-benzoquinone, and phenol are present in female full body extracts from *M. melolontha* and *M. hippocastani*, field experiments were performed addressing the question, whether swarming males discriminate between conspecific and heterospecific females. Males of both species prefer females when given the choice between females and males of the other species. However, they prefer conspecific females when females from both species are offered simultaneously. The results suggest that species-specific pheromone blends contribute to precopulatory reproductive isolation in sympatric populations of *M. melolontha* and *M. hippocastani*. But in contrast to findings in other sympatric scarab beetles, a blend emitted by forest cockchafer females is not a behavioural antagonist to European cockchafer males and vice versa. Furthermore, the respective blends are not indispensable prerequisites to find and select a mate, as it is the case in other insects.

10.1.2 Host plant volatiles

Responses of *M. melolontha* cockchafers to host plant volatiles were investigated both in the field and using electrophysiological techniques. Male cockchafers are attracted by volatiles from mechanically damaged leaves of *Fagus sylvatica* L., *Quercus robur* L., and *Carpinus betulus* L. Odours from intact *F. sylvatica* leaves are not attractive to *M. melolontha* males. In total, 16 typical plant volatiles are shown to elicit electrophysiological responses on cockchafer antennae, among them many green leaf volatiles typically emitted by damaged leaves. In the field the green leaf alcohols (*Z*)-3-hexen-1-ol, (*E*)-2-hexen-1-ol, and 1-hexanol attract males, whereas the corresponding aldehydes and acetates are behaviourally inactive. Thus, the function of sexual kairomones in the mate finding process of *M. melolontha*, can only be attributed to green leaf alcohols. Interestingly, the close relative, *M. hippocastani* responds only to (*Z*)-3-hexen-1-ol, not to the other leaf alcohols. Females are not attracted by any of the tested volatile sources.

To elucidate the structure–activity relationships of aliphatic alcohols, i.e., green leaf alcohols and non-natural analogues, both behavioural and physiological responses were studied in male and female *M. melolontha*. The compounds tested were saturated aliphatic alcohols with chain lengths between five and eight carbon atoms. Furthermore, the cockchafer's responses to six-carbon alcohols with (*E*)-2-, (*E*)-3-, (*Z*)-2-, (*Z*)-3-, and (*Z*)-4-configurated double bonds were tested. All compounds elicit dose-dependent responses on the antennae of both sexes. In general, males show a stronger normalized EAG response to the stimuli than females. In the field, only the naturally occurring six-carbon alcohols, i.e., 1-hexanol, (E)-2-, (Z)-3, and (E)-3-hexen-1-ol are attractive to *M. melolontha* males. The attractiveness depends on the molecules' structure. Females are not attracted by any of the tested compounds.

The results of the field and physiological experiments with natural and synthetic host plant volatiles highlight the function of the green leaf alcohols as sexual kairomones. No evidence was found that males or females use plant volatiles for host location.

10.1.3 Application ascpects

To optimize cockchafer lures, specific binary or ternary blends of (Z)-3-hexen-1-ol with phenol, and toluquinone or 1,4-benzoquinone, respectively have been tested in funnel trap experiments. In both species, *M. melolontha* and *M. hippocastani*, binary lures containing (Z)-3-hexen-1-ol combined with toluquinone or 1,4-benzoquinone, respectively, at a ratio of 10:1 are the most potent male attractants.

10.1.4 Oriented responses below ground

The larvae of the European cockchafer *M. melolontha* (white grubs) are polyphagous root feeders. The present laboratory study investigated whether white grubs orientate in the soil by chemical cues released from the roots. A new soil arena type was developed for this purpose. Former studies have shown that white grubs orientate along a CO_2 gradient. White grub responses to a gradient of synthetic CO_2 in the new soil arena corroborate these findings and demonstrate the suitability of the arena to investigate the orientation of scarab larvae. When roots from dandelion, a highly preferred host, and roots from red clover, an accepted host are offered, white grubs do not respond to the stimulus, even though the plant roots generate adequate CO_2 gradients. When synthetic CO_2 is offered as stimulus on the background of exudates extracted from undamaged dandelion roots, the white grubs do not respond as well. Thus, root chemicals most probably mask the white grubs' response to CO_2. However, future experiments addressing the role and the identification of the 'masking' compounds are needed to confirmed the hypothesis.

10.2 Conclusions

Results from the studies presented here unambiguously support the concept of sexual kairomones, as plant volatiles are the primary sex attractant in *M. melolontha*. However, it remains an open question whether other plant volatiles than GLV play a role in host finding and host selection.

The identified female beetle-derived male attractants are known as defensive compounds with antimicrobial activity. The presented results thus support the "secondary function hypothesis", according to which scarab beetle sex pheromones largely evolved from compounds used for defensive purposes. A specific blend of the sex pheromone toluquinone with the leaf alcohol (Z)-3-hexen-1-ol has been identified as the so far most powerful lure to cockchafer males.

The results show that species-specific pheromone blends contribute to, but do not warrant precopulatory reproductive isolation in sympatric populations of the European and the forest cockchafer. Hence, other mechanisms, which remain to be elucidated, may contribute to species isolation during the mate finding process.

Physiological experiments have shown that females perceive almost the same set of volatiles as males do. Females have been observed flying before the main swarming flight, when traps were already placed. A significant behavioural response to any offered stimulus has, however, never been detected. Nevertheless, these observations and the presented data do not allow conclusions about female olfactory orientation. Males locate females, which represent, as most traps do, spot-like odour sources. Females must find suitable host trees to feed on, and thus, may have a different search strategy. Future investigations regarding female cockchafer orientation need to take these considerations into account.

A new type of soil arena has been developed and proven to be suitable to investigate the orientation of white grubs. The presented results show that compounds emitted by the plant roots modulate the attractiveness of CO_2 to *M. melolontha* larvae. The exact function of these compounds and their chemical structure has to be elucidated in the future. The results highlight the need to intensify efforts to understand below-ground plant - herbivore interactions.

10.3 Deutsche Zusammenfassung

Feldmaikäfer, *Melolontha melolontha* L., sind in Gradationssphasen Schädlinge in Landwirtschaft und Gartenbau. Die polyphagen Larven (Engerlinge) fressen an Pflanzenwurzeln, während die Käfer Wirtsbäume vollständig entlauben können. Als Schädling wurde die Art in der Vergangenheit intensiv untersucht. Natürliche Duftstoffe, welche die Partnerfindung vermitteln oder dem Lokalisieren von Wirtspflanzen dienen, können in der Schädlingsbekämpfung eingesetzt werden. Nichtsdestotrotz war zu Beginn der hier vorgelegten Arbeit kaum etwas über die chemische Orientierung der Feldmaikäfer bekannt. Diese chemisch-ökologische Studie hatte zum Ziel, unter Einsatz chemischer Analytik sowie mit Hilfe von Freiland- und Laborexperimenten, Verhalten modifizierende Pheromone und Kairomone zu identifizieren.

10.3.1 Aufsuchen der Reproduktionspartner und Pheromonkommunikation

Maikäfer vollführen während der Abenddämmerung einen spektakulären Schwärmflug um die Kronen der Wirtsbäume. Zählungen der fliegenden und verharrenden Käfer ergaben, dass der Schwärmflug ausschließlich von Männchen ausgeführt wird, während Weibchen auf den Blättern, an denen sie fressen oder gefressen haben, sitzen bleiben. Im Freiland ausgeführte Landekäfigwahlversuche ergaben, dass Männchen Käfige, die Weibchen enthielten, gegenüber mit Männchen bestückten Käfigen bevorzugen. In gaschromatographischen Analysen mit gekoppelter elektroantennographischer Detektion wurden mehrere physiologisch aktive Verbindungen identifiziert, unter ihnen Toluchinon, Phenol und 1,4-Benzochinon, das Sexualpheromon des nahe verwandten Waldmaikäfers, *M. hippocastani* FABR. In Trichterfallenversuchen vermochte keines der Benzochinone, einzeln verwendet, Männchen anzulocken. Düfte mechanisch beschädigter Wirtsbaumblätter oder ein Mix sog. allgemeiner grüner Blattdüfte, der das Bouquet verletzter Blätter nachbildete, waren für Männchen jedoch hochattraktiv. Die Attraktivität des gleichen Duftbouquets wurde durch Zugabe von Toluchinon synergistisch gesteigert. 1,4-Benzochinon war hingegen nicht verhaltensaktiv. Somit wirken die Düfte beschädigter Blätter als Sexualkairomone, welche Männchen während des abendlichen Schwärmflugs zu Aufenthaltsorten der Weibchen führen. Das Sexualpheromon Toluchinon zeigt darüber hinaus die tatsächliche Präsenz eines arteigenen Weibchens an und erhöht in synergistischer Wechselwirkung die Attraktivität der Blattdüfte. Diese Befunde stellen den ersten Bericht über ein Sexualpheromon dar, das im Experiment über keine eigene Attraktivität verfügt, sondern die gleichzeitige Anwesenheit von Blattdüften erfordert, um seine Verhalten modifizierende Wirkung zu entfalten.

Phenol wurde sowohl in Ganzkörperextrakten von Feld- als auch von Waldmaikäferweibchen gefunden. In Fallenversuchen wirkte diese Verbindung auf Männchen beider Arten anlockend und erhöhte die Attraktivität des Blattalkohols (*Z*)-3-Hexen-1-ol. Bei Verwendung gleicher Anteile ist der kombinierte Duft von Phenol und (*Z*)-3-Hexen-1-ol für *M. hippocastani* Männchen weniger attraktiv als die Kombination von Phenol mit 1,4-Benzochinon. Für *M. melolontha* Männchen sind beide Duftkombinationen gleichermaßen attraktiv. Kombinierte Duftköder mit (*Z*)-3-Hexen-1-ol, Phenol und dem jeweiligen Benzochinon locken nicht mehr Männchen an als binäre Köder aus dem Blattalkohol und dem jeweiligen Benzochinon. Phenol ist somit ein weiterer Lockstoff für Männchen beider Arten. Wird Phenol mit optimierten Mischungen von

(*Z*)-3-Hexen-1-ol und dem jeweiligen Benzochinon (5 mg : 0.5 mg) kombiniert, gehen bei beiden Maikäferarten die Fangzahlen zurück. Daher wurde vom Begriff Sexualpheromon zur Charakterisierung dieser Verbindung abgesehen. Die genaue Funktion von Phenol ist noch aufzuklären.

Toluchinon, 1,4-Benzochinon und Phenol wurden in Ganzkörperextrakten von *M. melolontha* und *M. hippocastani* Weibchen gefunden. In Freilandexperimenten wurde daher der Frage nachgegangen, ob schwärmende Maikäfermännchen die Weibchen beider Arten anhand chemischer Stimuli unterscheiden. Feld- wie Waldmaikäfermännchen bevorzugen Weibchen, wenn ihnen die Wahl zwischen Männchen und Weibchen der jeweils anderen Art gegeben wird. Haben sie hingegen die Wahl zwischen arteigenen und artfremden Weibchen, werden die arteigenen Weibchen bevorzugt. Diese Ergebnisse weisen auf artspezifische Pheromonbouquets hin, welche zur präkopulatorischen reproduktiven Isolation der Arten in sympatrischen Populationen beitragen. Im Gegensatz zu Befunden bei anderen Scarabaeidenarten, ist im Pheromonbouquet eines Waldmaikäferweibchens aber kein Verhaltensantagonist für Feldmaikäfermännchen enthalten und umgekehrt. Ebenso sind die arteigenen Pheromonbouquets offenbar keine unabdingbare Vorraussetzung zur Partnerfindung und Partnerwahl, wie es bei anderen Insekten der Fall ist.

10.3.2 Wirtspflanzendüfte

Die Reaktionen von adulten *M. melolontha* auf Wirtspflanzendüfte wurden sowohl elektrophysiologisch als auch im Freiland untersucht. Feldmaikäfermännchen wurden von den Düften mechanisch beschädigter Blätter von *Fagus sylvatica* L., *Quercus robur* L., und *Carpinus betulus* L. angelockt. Intakte *F. sylvatica*-Blätter waren nicht attraktiv. Maikäferantennen nehmen mindestens 16 charakteristische Blattdüfte, unter ihnen zahlreiche sog. allgemeine grüne Blattdüfte (GLV), wahr. In Freilandversuchen erwiesen sich die Blattalkohole (*Z*)-3-hexen-1-ol, (*E*)-2-hexen-1-ol und 1-Hexanol als attraktiv, während die homologen Aldehyde und Azetate nicht attraktiv waren. Somit kann nur den Blattalkoholen die Funktion von Sexualkairomonen bei der Partnerfindung zugewiesen werden. Interessanterweise wird der nahe verwandte Waldmaikäfer ausschließlich durch (*Z*)-3-Hexen-1-ol, nicht jedoch durch die anderen Blattalkohole angelockt. Keiner der angebotenen Düfte wirkte auf Weibchen attraktiv.

In elektrophysiologischen und Verhaltensexperimenten wurden mit natürlich vorkommenden Blattalkoholen und natürlich nicht vorkommenden Analoga Struktur-Aktivitäts-Beziehungen untersucht. Es kamen gesättigte aliphatische Alkohole mit fünf bis acht Kohlenstoffatomen sowie einfach ungesättigte Alkohole mit sechs Kohlenstoffatomen und der Doppelbindung in (*E*)-2-, (*E*)-3-, (*Z*)-2-, (*Z*)-3-, und (*Z*)-4-Konfiguration zum Einsatz. Alle getesteten Verbindungen riefen dosisabhängige Reaktionen der Antennen hervor. Die standardisierten Reaktionen von Männchenantennen waren im Allgemeinen stärker als die von Weibchenantennen. Im Freiland waren ausschließlich die natürlich vorkommenden Blattalkohole 1-Hexanol, (*E*)-2-, (*E*)-3- und (*Z*)-3-Hexen-1-ol attraktiv. Der Grad der Attraktivität hing von der Molekülstruktur ab. Weibchen wurden von keinem getesteten Blattalkohol angelockt.

Die Ergebnisse dieser Freiland- und Laborexperimente unterstreichen die Funktion von Blattalkoholen als Sexualkairomone. Es wurden keine Hinweise gefunden, dass Männchen oder Weibchen Blattdüfte bei der Wirtsfindung nutzen.

10.3.3 Anwendungsaspekte

Zur Optimierung von Duftködern wurden binäre und ternäre Duftgemische mit den Bestandteilen (Z)-3-Hexen-1-ol, Phenol und dem jeweiligen Benzochinon in Trichterfallenversuchen in Feld- und Waldmaikäferfluggebieten getestet. Sowohl *M. melolontha* als auch *M. hippocastani*-Männchen werden am stärksten von binären Duftködern mit (Z)-3-Hexen-1-ol und Toluchinon bzw. 1,4-Benzochinon im Verhältnis 10:1 angelockt.

10.3.4 Orientierung im Boden

Die Larven des Feldmaikäfers (Engerlinge) ernähren sich polyphag von Pflanzenwurzeln. Im Rahmen dieser Laborstudie wurde untersucht, ob sich Engerlinge anhand von wurzelbürtigen chemischen Stimuli orientieren. Zu diesem Zweck wurde eine neuartige Bodenarena entwickelt, die es ermöglicht unbeschädigte Wurzeln intakter Pflanzen als Stimulus anzubieten. Frühere Studien kamen zu dem Ergebnis, dass Engerlinge sich entlang von Kohlendioxidgradienten (CO_2) orientieren. Nach Aufbau eines technischen CO_2-Gradienten, orientierten sich die Engerlinge zur Quelle hin. Somit wurden die früheren Befunde bestätigt, aber vor allem die Eignung der Bodenarena für Orientierungsversuche an Scarabaeidenlarven nachgewiesen. Wenn als Ernährungsressource hochpräferierte Löwenzahnwurzeln oder als Ressource angenommene Rotkleewurzeln als Stimulus angeboten wurden, erfolgte keine zum Stimulus gerichtete Orientierung, obwohl adäquate CO_2-Gradienten von den Wurzeln erzeugt wurden. Wenn erneut synthetisches CO_2, dieses Mal aber vor dem Hintergrund von Exsudaten intakter Wurzeln angeboten wurde, erfolgte ebenfalls keine Orientierung zur CO_2-Quelle hin. Diese Befunde deuten darauf hin, dass von den Wurzeln abgegebene Verbindungen eine Verhaltensantwort auf den CO_2-Gradienten unterbinden oder anders ausgedrückt, die Attraktivität des CO_2 ‚maskieren'. Allerdings muss diese Hypothese durch Aufklärung der chemischen Struktur der für die Wirkung verantwortlichen Verbindungen in zukünftigen Untersuchungen untermauert werden.

10.4 Schlussfolgerungen

Blattalkohole wurden als primäre Lockstoffe bei der Partnerfindung von *M. melolontha* nachgewiesen. Die Ergebnisse dieser Studie unterstützen somit unzweifelhaft das Konzept der Sexualkairomone. Es bleibt allerdings eine offene Frage, ob andere Pflanzendüfte bei der Wirtsfindung eine Rolle spielen.

Die in Weibchen nachgewiesenen, Männchen anlockenden Verbindungen sind aus anderen Taxa als Abwehrstoffe bekannt und verfügen über antimikrobielle Aktivität. Ihre Funktion im Rahmen des Reproduktionsverhaltens unterstützt die „secondary function hypothesis". Demnach sind „Sexualpheromone" bei Scarabaeiden vielfach Verbindungen mit ursprünglich defensivem Charakter. Ein spezifisches Mischungsverhältnis von (Z)-3-Hexen-1-ol mit den Sexualpheromonen von *M. melolontha* und *M. hippocastani* wurde als bislang attraktivste Duftkombination nachgewiesen.

Nach den vorliegenden Ergebnissen tragen artspezifische Pheromonbouquets zur präkopulatorischen reproduktiven Isolation in sympatrischen Populationen der untersuchten Arten bei, gewährleisten diese aber nicht. Andere Mechanismen sollten daher zur reproduktiven Isolation beitragen und bedürfen der Aufklärung.

In elektrophysiologischen Experimenten wurde nachgewiesen, dass Weibchen etwa die gleichen Düfte wahrnehmen wie Männchen. Trotzdem wurde zu keinem der in dieser Untersuchung angebotenen Stimuli eine Verhaltensreaktion nachgewiesen. Allerdings lässt dieser Befund keine Rückschlüsse auf die olfaktorische Orientierung der Weibchen zu. Männchen müssen Weibchen lokalisieren, die ebenso wie gängige Fallentypen punktförmige Duftquellen darstellen. Zum Aufsuchen von Wirtsbäumen oder Eiablageflächen könnten Weibchen hingegen eine andere, mehr auf Flächen bezogene, aber durchaus olfaktorisch geleitete Suchstrategie entwickelt haben. Diese Überlegungen sollten in zukünftige Versuche zur Orientierung von Weibchen einfließen.

Es wurde eine neuartige Bodenarena entwickelt und getestet. Die Ergebnisse zeigen, dass von Wurzeln abgegebene Verbindungen die Attraktivität von CO_2 auf Engerlinge modulieren. Die exakte Funktion dieser Verbindungen und ihre Identität müssen in zukünftigen Arbeiten untersucht werden. Die Befunde zeigen aber vor allem, dass die Interaktionen von Pflanzen und Herbivoren im Boden mit noch größeren Anstrengungen als bislang untersucht werden müssen.

Danksagung

Diese Arbeit ist nur durch die Mitwirkung zahlreicher Menschen innerhalb wie außerhalb der Arbeitsgruppe Angewandte Zoologie/Ökologie der Tiere möglich geworden.

Ganz besonders möchte ich meiner Betreuerin, Monika Hilker, für die große Unterstützung, die zahlreichen Anregungen und das entgegengebrachte Vertrauen danken.

Besonderer Dank gilt ebenfalls Joachim Ruther, der mich in das Maikäferprojekt eingeführt hat und immer wieder mit Rat und Tat bereit stand.

Frank Müller hat mit seinem Erfahrungsschatz und großer Lauffreude geholfen, so manch technisches Problem zu bewältigen.

Ute Braun hat sich um die Elektrophysiologie und im Verein mit Renate Jonas um das Wohl der Engerlinge große Verdienste erworben. Renate Jonas gebührt ein besonders herzliches Dankeschön für den ‚nicht wissenschaftlichen', aber unentbehrlichen Anteil am Gelingen dieser Arbeit.

Vielen Dank auch an Urte Kohlhoff, der Lotsin in den Klippen der Bürokratie.

Torsten Meiners hatte immer ein offenes Ohr und wo erforderlich einen guten und hilfreichen Rat.

Alle AG-Mitglieder haben Anteil an der guten und freundschaftlichen Arbeitsatmosphäre. Euch allen ganz herzlichen Dank.

Außerhalb der Arbeitsgruppe haben zahlreiche Mitarbeiter von Pflanzenschutzämtern, Förster, Landwirte, Ortsvorsteher u.a. die Versuche im Freiland ermöglicht. Ganz herzlicher Dank geht an Manfred Fröschle, Jean-Luc Houot, Robert Mougin, Vincent Potaufeux, Familie Bumen, Horst Gossenauer-Marohn, Kerstin Jung, Winfried Schüler, Dagmar Leisten, Franz Späth, Robert Weiland, Nico Versch, Willi Georg Muth, Otto Fäth, Ulrich Benker, Monika Rehm sowie die studentischen Hilfskräfte Tanja Bloss, Martha Carboni, Nana Hesler, Jana Collatz, Nadine Herrmann, Janina Lehrke, Maya Ulbricht und die promovierte Hilfskraft Stefan Sieben. Mögen mir die verzeihen, die zu erwähnen ich vergessen habe.

Ganz besonderer Dank gilt meiner Familie – mir fehlen die Worte.